The Global Life of Mines

DISLOCATIONS

General Editor:
Don Kalb, *University of Utrecht & Central European University*

The Dislocations Series inquires about the relational, spatial, and temporal dimensions of power under globalized capitalism. Through stimulating critical perspectives and cross-disciplinary frameworks, the series provides a forum for politically engaged, ethnographically informed and theoretically incisive responses.

Recent titles:

Volume 37
The Global Life of Mines: Mining and Post-Mining in Comparative Perspective
Edited by Antonio Maria Pusceddu and Filippo M. Zerilli

Volume 36
The Making and Unmaking of the Ukrainian Working Class: Everyday Politics and Moral Economy in a Post-Soviet City
Denys Gorbach

Volume 35
Insidious Capital: Frontlines of Value at the End of a Global Cycle
Edited by Don Kalb

Volume 34
From Village Commons to Public Goods: Graduated Provision in Urbanizing China
Anne-Christine Trémon

Volume 33
Corporate Social Responsibility and the Paradoxes of State Capitalism: Ethnographies of Norwegian Energy and Extraction Businesses Abroad
Edited by Ståle Knudsen

Volume 32
Glimpses of Hope: The Rise of Industrial Labor at the Urban Margins of Nepal
Michael Hoffmann

Volume 31
Bulldozer Capitalism: Accumulation, Ruination, and Dispossession in Northeastern Turkey
Erdem Evren

Volume 30
Facing the Crisis: Ethnographies of Work in Italian Industrial Capitalism
Edited by Fulvia D'Aloisio and Simone Ghezzi

Volume 29
Big Capital in an Unequal World: The Micropolitics of Wealth in Pakistan
Rosita Armytage

Volume 28
Fifty Years of Peasant Wars in Latin America
Edited by Leigh Binford, Lesley Gill and Steve Striffler

For a full volume listing, please see the series page on our website: https://www.berghahnbooks.com/series/dislocations

The Global Life of Mines

Mining and Post-Mining in Comparative Perspective

Edited by Antonio Maria Pusceddu
and Filippo M. Zerilli

berghahn
NEW YORK • OXFORD
www.berghahnbooks.com

First published in 2024 by
Berghahn Books
www.berghahnbooks.com

Library of Congress Cataloging-in-Publication Data
Names: Pusceddu, Antonio Maria, editor. | Zerilli, Filippo M., editor.
Title: The global life of mines : mining and post-mining in comparative perspective / Antonio Maria Pusceddu, Filippo M. Zerilli.
Description: First edition. | New York : Berghahn, 2024. | Series: Dislocations ; v. 37 | Includes bibliographical references and index.
Identifiers: LCCN 2024004312 (print) | LCCN 2024004313 (ebook) | ISBN 9781805395911 (hardback) | ISBN 9781805395928 (epub) | ISBN 9781805395935 (web pdf)
Subjects: LCSH: Mineral industries--Social aspects. | Abandoned mines--Social aspects. | Mines and mineral resources--History. | Industrial archaeology.
Classification: LCC HD9506.A2 G57 2024 (print) | LCC HD9506.A2 (ebook) | DDC 338.2--dc23/eng/20240412
LC record available at https://lccn.loc.gov/2024004312
LC ebook record available at https://lccn.loc.gov/2024004313

British Library Cataloguing in Publication Data
A catalogue record for this book is available from the British Library

ISBN 978-1-80539-591-1 hardback
ISBN 978-1-80539-592-8 epub
ISBN 978-1-80539-593-5 web pdf

https://doi.org/ 10.3167/9781805395911

CONTENTS

ILLUSTRATIONS

INTRODUCTION

Timescapes of Extraction

Antonio Maria Pusceddu and Filippo M. Zerilli

Prelude

In 1982, Eric R. Wolf published his magnum opus *Europe and the People Without History*. This long-length book included 'cartographic illustrations' selected by Noël L. Diaz. Two copper engravings by the sixteenth-century Dutch engraver and publisher Theodore de Bry open the first two parts of the book ('Connections' and 'In Search of Wealth'). The second of these two engravings is titled 'Washing gold' (Wolf 1997 [1982]: 126–27). To the right, half-naked men dig the gallery of a gold mine with picks, while other men – we glimpse them in the darkness – are timbering the galleries. Still others run with buckets of gold on their heads, and some tip gold at the foot of a finely-dressed European lord, who is sat in a chair. He holds a rod and is surrounded by men with spears and swords. Far distant, other armed men oversee the mining activities.

The engraver Theodore de Bry (1528–1598) was well-known for his illustrations of the European colonization of the Americas and the imaginative representations of Amerindian populations, even though he had never seen them first hand, relying on second-hand descriptions, such as travel accounts and reports by those who had been to the Americas and saw the 'new' lands (Van Groesen 2019).[1] To represent gold mining, Theodore de Bry most likely resorted to his knowledge of mining activities in Europe. In 1525 it was reported that the European metallurgical sector employed a massive 'one hundred thousand workers' (Moore 2007: 127). Likely, the bodies and toiling of mineworkers in Europe could have served as a model to depict the exploitation of indigenous and enslaved African labourers. As Europe was flooded with the silver extracted in the Americas, 'tripling the European supply' between 1503 and 1660 (Wolf 1997 [1982]: 139), silver mines on the old continent were exposed to increased competition, which eventually led to reduced extraction and therefore a reduced workforce (Komlosy 2018: 126). To supply Europe with American silver, mineworkers were largely replaced with the cheap labour of Amerindian communities and by enslaved labourers traded from the West African coast across the Atlantic.

Theodore de Bry's engraving captures a meaningful moment at the dawn of modern capitalist globalization. The European conquest and plunder of the Americas ushered in a new era of predatory extraction, employing exploitation and violence on an unprecedented scale. Mining laid the foundations for a transatlantic system of wealth accumulation through the violent dispossession of the Amerindian peoples and the environment. On the eve of colonial conquest, the rapid expansion of mining in Europe sparked major peasant uprisings in response to the enclosure of communal forests and the environmental destruction of livelihood means (Moore 2007). From the outset, the modern world system was built on the sustained expansion of extractive activities, on new logics of labour and nature exploitation that paved the way for the rise of a new regime of socioecological relations (Machado Aráoz 2018 [2014]; Moore 2007, 2015).

Five centuries after De Bry had visually imagined gold mining in the Americas, resource extraction has grown dramatically, providing the indispensable raw materials to produce every kind of commodity that pervades our daily lives; as Jacka (2018: 62) puts it, 'our entire livelihoods are utterly dependent on minerals'.[2] For at least two centuries now, resource extraction has profoundly altered entire ecosystems and ways of life, generating lasting pollution problems that are increasingly difficult to manage (Bridge 2004; Jacka 2018: 66–68). Resource extraction has also changed and shaped distinctive minescapes and social identities, through the ways of mobilizing and organizing labour. The expansive and pervasive nature of resource extraction is now also reflected in the growing literature on mining. Since the beginning of the new millennium, interdisciplinary research on resource extraction has literally boomed.[3] This emerging field of study brings together the work of anthropologists, critical geographers, archaeologists, sociologists and environmental historians (at least since the work of Knapp, Pigott and Herbert 1998). Different sensibilities contribute to making the study of resource extraction a truly interdisciplinary field that fruitfully intersects with other similar interdisciplinary endeavours, from deindustrialization studies to political ecology.

This book aims to contribute to this growing literature by putting together mining and post-mining in a global perspective; that is, by approaching the spatial and temporal articulations of mining and post-mining as interrelated and coexisting features of the same global minescape. The main point we want to make is that placing mining and post-mining in a global perspective helps us account for the interconnections between the diverse and diversified worlds of resource extraction and their lasting effects over time, and to think about their legacies as intimately connected to the contemporary extractive geographies. To do this, we propose framing mining and post-mining sites through the idea of the 'global life of mines', which we conceive as a methodological framework for addressing the variations and permanence of resource extraction and its aftermath. Taken together, the chapters in this book offer insights into the multiple temporalities entangled in the timescapes of resource extraction (D'Angelo and Pijpers 2018a, 2018b) and how such temporalities, far from being captured by conventional linear chronologies (of clocks, calendars, calculations, estimates), coexist and interact in producing the global minescape. Thinking about mining and post-mining in terms of their 'global life' means to combine the spatialities of dispersed sites of (post)extraction with a timescape perspective that highlights the temporal intricacy and multifaceted temporal dimensions of socio-environmental life (Adam 1998). Consequently, the 'global life of mines' entails understanding extractive spaces through time and the temporalities of extraction through space.

This book brings into conversation researchers who address different dimensions, implications and temporalities of resource extraction and its aftermath.[4] In this Introduction, we highlight a 'global' outlook as a productive way of thinking about extractive activities and their afterlives. We clarify the meaning of the global, what is global in the life of mines, and how we believe it is crucial to bring together mining and post-mining in a comparative perspective. We think of the global life of mines as a comprehensive framework for thinking about circuits of extraction, mining sites and their exhausted landscapes as intimately interconnected.

Timescapes of Extraction

The debate on the Anthropocene as a proper term to identify a new geological epoch, characterized by the growing impact of human activities on earth (Crutzen and Störmer 2000), stimulated the proliferation of alternative or complementary characterizations with the suffix 'cene': from Capitalocene to Plantationocene, from Chthulucene to Plasticocene, Wasteocene, etc.[5] Whatever term we may opt for (and we may even opt for more than one or all at once), we cannot escape the fact that we are living – more than ever – in 'the mineral age' (Jacka 2018). In recent decades, mineral extraction has reached an unprecedented scale and intensity – and it is expected to grow further. The expanding electric vehicle industry, 'clean' energy technologies (from wind turbines to solar panels) and the widespread digital technologies are devouring increasingly enormous quantities of minerals (Geenen, this book). A report by the International Energy Agency shows how mineral-intensive these technologies are, compared to fossil fuel technologies: 'A typical electric car requires six times the mineral inputs of a conventional car', while 'an onshore wind plant requires nine times more mineral resources than a gas-fired plant of the same capacity' (IEA 2021). Considering this ratio, the IEA estimates that reaching the goals of the Paris Agreement by 2040 (that is, to limit global warming to below 2 degrees Celsius) would require four times the current mineral supply for 'clean' energy technologies. The increasing demand for minerals is especially noteworthy for lithium and graphite, truly 'timely assets' (Ferry and Limbert 2008).[6] According to the World Bank (2020), to meet the goals of the Paris Agreement by 2050, 'production would need to ramp up by nearly 500 percent'.

The mineral boom was already on its way at the end of the last century (Ballard and Banks 2003). Over the last fifty years, the 'global annual volume of extracted materials has tripled' (D'Angelo and Pijpers 2022b: 2). The new millennium has seen an impressive expansion of extractive activities, largely due to the liberalization of the mining sector in developing countries (Jacka 2018: 63) – a crucial factor in post-Cold War neoliberal worldwide deregulation (Kirsch 2014). African countries witnessed an unprecedented rise in foreign investments in the extractive sector – a fact that contributed to the expansion of a flexible workforce in the mining industry (Rubbers 2021). Regions targeted for centuries by European extractivist exploitation started to pursue neo-extractivist policies by attracting corporate investments and making the mining sector a pillar of export-oriented growth policies (Arboleda 2020).[7] Latin American countries – like Bolivia (Ødegaard, this book) – have been trying to recast mining activities as a key driver for the development of a nationally-owned industry (Barandiarán 2019; Revette 2016). In 2008 the European Union launched the Raw Materials Initiative (RMI), intending to support new mining projects in Europe. The worldwide bustling expansion of mineral extraction is also marked by an incredible variation regarding types of mining operations, technological levels and managing patterns and work relations, ranging from technologically advanced industrial large-scale mining (ILM) to the myriad of artisanal and small-scale mining (ASM) operations. We are thus witnessing something very old and, at the same time, something novel, for its scale, extension and impact. Extractive activities have been around as long as human civilizations. The early days of capitalist development have seen the rapid expansion of profit-seeking extractive activities. Colonialism triggered the formation of modern extractive geographies and racialized labour forces, while the take-off of industrial capitalism in the nineteenth century generated a growing demand for coal and minerals (lead and zinc, for instance).

The making of new frontiers of extraction has often entailed the demise of existing ones, thus leaving behind derelict and exhausted extractive landscapes (Calvão 2011; Oakley 2018). As early as the sixteenth century, the abundant inflow of silver and other minerals from the Americas badly affected European mining regions. The latter expanded again in the nineteenth century, shaping a combative and iconic labour force, frequently racialized and demonized, such as in rapidly industrializing Britain (Metcalfe 1990; Mitchell 2013: 21–42; Thompson 1963). The decline of the European mining industry,[8] beginning in the 1960s, was paralleled by the rapid expansion of the mining industry across the African continent, far beyond the one already established in the Copperbelt (Powdermaker 1962).[9] Since the 1960s, other resource frontiers have emerged, increasingly targeted by the unbridled circulation of capital – such as Papua New Guinea[10] or other areas of Southeast Asia (Ballard and Banks 2003).

The making of resource frontiers is one of the connecting threads between the chapters in this book, which examine the constant dialectic between waste and value (Gidwani 2012, 2013), emphasizing the colonial legacies of capitalist appropriation of devalued ('cheap', following Moore 2015; Patel and Moore 2018) nature and of socioecological dispossession. Resource-rich territories are therefore appropriated through the 'discovery paradigm' (Burton 1997; Kirsch 2014) as 'virgin land' (White, this book), 'terra nullius' (Mesbah, this book) or through the narratives of valorization and economic development of 'unused' (hence wasted) resources (Buu-Sao, this book). The concept of 'resource frontier' lies at the intersection between temporal and spatial scales (Jacka 2018; Tsing 2005: 28); while it is a central concept to think about how regions are turned into territories of extraction, and about the powerful boundary work of resource-making processes (Ferry and Limbert 2008), it is also relevant to explore how post-mining areas are being re-envisioned as new frontiers. As shown by Ødegaard (this book), the Norwegian state-led project of 'restoring to nature' the coal mining Arctic Svalbard archipelago mobilizes the narrative of a post-carbon socio-environmental future to turn a resource frontier into 'nature's frontier'; a shift that toys with visions of new subjects for a new era, from which coal miners are excluded as relics of a bygone age. Buu-Sao (this book) examines how corporate powers mobilize the narrative of sustainable mining (a 'corporate oxymoron' analysed by Kirsch 2010) to re-create the post-mining Andalusian region as a brand new technological frontier of resource extraction. Such operations rely on the 'strategic mobilization of temporal frames' and the 'historical amnesia' of the mining past (Luning 2012).

The Lives and Afterlives of Extraction

As we acknowledge the relevance and impact of mining activities in modern capitalist development, we can also think about the afterlives of extraction as the disseminated geographies of capitalist dissipation, depletion and exhaustion; and as lasting marks of socio-environmental change and capital's temporal alienation.[11] As more areas of the world are being turned into new resource frontiers, older mining regions are burdened with the long-standing problematic socio-environmental legacies of resource extraction and mineral processing. Few human activities move more volumes of earth than mining. Mining can profoundly alter the morphology and hydrology of extensive territories, affecting the flora and fauna as well as subsistence practices and the sociocultural maps of indigenous life. Mineral processing technologies – even the most rudimental – make use of chemicals (mercury, cyanide etc) that spill into creeks and rivers, causing the contamination of soils and waters (Bridge 2004; Mesbah, this book). The environmental impact of mining activities – especially large-scale industrial extraction – is not only lasting but also extremely costly, either in terms of remediation or in terms of any necessary measures to limit further damages, such as infrastructure or dams to prevent spillover from highly contaminated drainage areas. Such costs are hardly borne by the mining companies that are responsible for them. The socio-environmental costs of mining are externalities that do not count in production costs; they are passed on to the population and public authorities, often leading to exhausting negotiation processes for obtaining compensation. For instance, in the United States it has been estimated that 'more than 156 abandoned hard rock mining sites have been targeted for federal clean up, an intervention that will cost the US government an estimated $15 billion, more than ten times the annual Superfund budget for all environmental disasters' (Kirsch 2014: 4).

Post-mining regions have also to face the socio-economic consequences of mine closures. Unemployment and economic downturn are the most immediate and pressing concerns, often resulting in the depopulation of entire areas and the fragmentation of formerly cohesive communities, turning them into spectral memories of the past (Kideckel 2018). Similarly, post-mining regions may suffer from cultural disorientation, resulting from perceptions of disposability and the severe crisis of working-class cultures (Strangleman, Rhodes and Linkon 2013). When this happens, it can take a long time to deal with the afterlives of mining activities, in terms of finding ways and means to tackle job losses and socio-environmental degradation and also negotiate proper compensatory agreements, or pursue meaningful collective memories of the past. The case analysed by Zanini (this book) reveals the historical othering of mineworkers within Alpine communities and how othering persists in the friction over the preservation of mining memories. Following the closure of goldmines in the 1960s, mineworkers migrated overseas (e.g. to Uganda) to seek employment in the mining sectors; meanwhile, both the centuries-old mining history and the painful memory of miners' toil were marginalized in the touristic reinvention of the traditional mountain community. Other times, the aftermath of mining activities can be so dire and hard (and costly) to come to terms with that the continuing 'slow violence' (Nixon 2011) of these legacies is ignored, intersecting with other forms of racial and social oppression, compounding the devaluation of indigenous and minority populations, stigmatized as worthless (Mesbah, this book; White, this book). The post-mining transition in the Svalbard archipelago, reimagined by the state as a laboratory of environmental restoration, displaces 'the miner' from the new narrative of post-extraction; mineworkers come to represent the negative temporal other of the new domain 're-created as a post-extractivist and environmentalist showcase' (Ødegaard, this book). In the mining revival of the Andalusian wastelands, the historical figure of the mineworker is overshadowed by the narrative of technologically advanced and sustainable mining operations, smart jobs and highly skilled technicians (Buu-Sao, this book; see Lanzano 2022). The long political history of Andalusian mineworkers has been monumentalized as a thing of the past. All these diversified histories of dispossession make up the uneven geography of mining and post-mining.

Operating mines and abandoned wastelands are connected through the same logic of accumulation and profitability; as such, they can be viewed as spatial reflections of the multiple (and stratified) temporalities of capital (Tomba 2009). By bringing mining and post-mining together we are also putting into conversation spatial and temporal notions of resource extraction, old mining sites and new territories of extraction (Luning 2018: 281–82). As previously emphasized, the global life of mines is as much about time as it is about space. Recent anthropological scholarship has provided insightful approaches to untangling the multiple mining temporalities (D'Angelo and Pijpers 2018b). Drawing on fieldwork on artisanal gold mining in Sierra Leone, Engwicht (2018) underlines the significant role of miners' concept of the future and their expectations for a better life, while the work of Askland (2018) in the coal mine region of New South Wales, Australia, approaches the temporality of mining through the lens of (im)mobility, showing how mining-induced displacement parallels involuntary immobility, producing what she calls 'broken time'. While mining projects polarize labour spatial mobility (Castillo and Brereton 2018), the closure or abandonment of mining sites highlights specific temporal strategies and dimensions, as shown by Oakley (2018) in a comparative analysis of several mining sites transformed into heritage, tourist destinations. Similarly, the chapters in this book focus on specific dimensions of the temporalities embedded in the end or suspension of extractive activities – and their legacies. It is not only about the afterlives of resource extraction – its continuing and long-standing consequences – but also about the history of resource extraction as a persistent and indelible presence. The authors explore the multiple forms and shapes of this absent presence – from the tragic and deadly embodiment of uranium mining (Mesbah, this book) to the reimagination of mining as a contested heritage (De Souza Santos; Zanini, this book); from the haunting spirits of dead enslaved labourers in the Brazilian gold mines (De Souza Santos, this book) to the silenced histories of evicted Timbisha Shoshone families in California's Death Valley (White, this book). The book contributes to the current burgeoning debates on resource extraction by proposing a possible framework for thinking about the temporality of resource extraction as already intimately connected with its afterlives, its consequences and legacies. The chapters show how resource extraction already contains the problematic temporality and materiality of its afterlives. In fact, both concepts – 'resource' and 'materiality' – need to be understood and refined on the ground, according to specific mining and post-mining timescapes. Theorizing how resources are entangled with experiences and politics of time, Ferry and Limbert (2008) unpacked the conventional notion of 'resources' and explored how practices of resource-making and resource-claiming are made and – at the same time – produce temporal imaginings (see Buu-Sao; Ødegaard, this book). In this vein, comparing how water and oil refer to different historical or fictional narratives in Oman, Limbert (2016) reveals how certain natural resources might embody or incorporate specific events, processes and temporal dimensions, disclosing certain environmental futures (see Mesbah, this book). However, as Limbert suggests, the meaning, management and use of 'natural' resources from the soil is never given but always subject to negotiations (that is, socially and politically constructed), as they represent a source of tension and potential conflict, revealing unequal distribution of wealth, power disparities and even different ontologies. From this perspective, the conventional idea and the concept of 'materiality' itself should not be taken for granted. Actually, the materiality of resources also has a relational character within specific socio-historical environments in which they are generated and transformed, within and beyond the capitalist market (Richardson and Weszkalnys 2014). Nevertheless, the environmental history of modern resource extraction is a history of continuing and persistent human displacement and socioecological degradation (Bridge 2004; Kirsch 2010; Mesbah, this book; White, this book).

Global Minescapes

There is little doubt that 'resource extraction is more than ever a global, planetary affair' (D'Angelo and Pijpers 2022b: 2). Yet, at the beginning of the twentieth century, Ballard and Banks (2003: 287) pointed to the paradox of studies being 'persistently parochial and regional in their scope' in spite of 'the transnational nature' of the mining industry. Their benchmark reference was Godoy's previous overview of the anthropology of mining, which relied on a relatively limited number of studies (Godoy 1985). Decades of profound transformation separate Godoy's in-depth review from the one by Ballard and Banks. Neoliberalism thrived with the emblematic defeat of British mineworkers by Margaret Thatcher. Deindustrialization advanced steadily in Western countries (High 2021). Industrialization flourished in Asia, with China developing a peculiar model of socialist-minded capitalism. The former African colonies – now formally independent states – were struggling with overwhelming debt from World Bank development loans. The collapse of the Soviet bloc marked a significant global event. The three-worlds division gave way to the ambiguous dichotomy between the Global North and the Global South. The globalized world was undergoing profound changes. Ballard and Banks wrote amidst burgeoning discourses of globalization that emphasized the transnational flows and networks of a borderless world, unified by the proliferation of information technologies. However, despite the sense of immateriality conveyed by the enthusiastic hailing of the rising digital world, none of this would have been possible without the thousands of workers digging craters, scraping away soil and operating sophisticated technologies to extract mineral wealth from the depths of the earth.

The idea of connection and interrelation is inherent in the very meaning of the word global, at least in the ways it has been widely used in a variety of domains since the late twentieth century (e.g. Conrad 2016; Tsing 2005; Van der Linden 2008). The idea of interconnection is so powerful that 'disconnection' has been theorized as the unsettling consequence of modern deindustrialization (Ferguson 1999; Vaccaro, Harper and Murray 2017). The inherent connectedness of the 'global' underlies the emergence of new geopolitical categories. The Global South, for instance, is not a simple substitutive term for former Third World nations, but it entails the proactivity of those countries (or some of them) as an interrelated whole in the face of the Global North (Kalb and Steur 2015). Theorizations of a 'global ethnography' (Burawoy et al. 2000) point to the connections between micro-practices and macro-structures, rethinking research outlooks on multiscalar terms, to grasp the threads connecting specific locales to wider processes. Likewise, the exploration of ethnographic locales becomes 'an ethnography of global connections' (Tsing 2005). Two decades earlier, Wolf (1997 [1982]: 3) made the central assertion that 'the world of humankind constitutes a manifold, a totality of interconnected processes', emphasizing the challenge of explaining such connections rather than merely describing them, and how these coexist as an expression of the field of forces shaping the more conventionally understood 'ethnographic site'.

While the global perspective can offer valuable insight into the connectedness of disparate geographies and stories of resource extraction, it is important to discuss its relevance alongside two other fundamental perspectives that emphasize the interconnection of historical and ecological processes. One is the world-system perspective, and the other is the planetary perspective. Immanuel Wallerstein's (1974) world-system theory represents the main theoretical contribution to the theorization of the systemic interconnection of historical processes since the worldwide expansion of historical capitalism in the sixteenth century. The world-system consists of the relations between the core, the semi-periphery and the periphery in the spatial organization of capital accumulation. The historical definition of these geographies of asymmetrical relations in the making of the modern world has undergone shifts and changes, without substantially altering the polarization between core regions accumulating surplus extracted from the peripheries.[12] The world-system theory provides a comprehensive explanation of the uneven geographies of labour and wealth accumulation. Despite its ability to provide a framework for understanding large-scale structural mechanisms, it falls short when it comes to accounting for the specific spatial and historical differences in centre-periphery relations (Wolf 1997 [1982]: 23). Methodologically, we need to turn to a global perspective to consider these variations as part of a wider historical process, without relinquishing the differences and variations that ethnographic investigations can bring up and disclose. The global perspective can therefore be understood as a methodological sensibility to make sense of variations and differences in the unfolding of world-system relations. Adopting a global perspective to look at the ethnographies and geographies of resource extraction and post-extraction gathered in this book, we aim at tracing their stories and forms of connection to (and 'disconnection' from) larger historical dynamics at work in the production of diversified minescapes.

The planetary perspective has recently been advocated as a more appropriate approach for thinking about the contemporary dimension of resource extraction (Arboleda 2020; Labban 2014). Labban (2014) suggests that mining 'should be considered planetary in scale' and detached from the territorial constraints of traditional mining – constraints dictated by the localization of metal ores. The planetary mine is therefore to be understood as the full-scale extension of metal extraction through the combination of hyper-sophisticated technologies that make it possible to extract metal from waste – regardless of their location. As such, the contemporary geographies and infrastructures of extraction are 'vastly dispersed across space' and entangled in a global apparatus of production and exchange, in ways that supersede the conventional world-system view of core-periphery relations (Arboleda 2020: 5).

Following labour historian Van der Linden (2012), it is worth noting that a global perspective may not always entail a theorization of the global systemic interconnection of phenomena, nor does it evade the problematic relevance of resource extraction in its planetary dimension, the intricate networked circuits of capital, or the overall alterations of earthly ecological systems. Putting things in a global perspective means thinking about localized phenomena through the lens of broader multiscalar processes. As a methodological posture, the global perspective assumes that we can better understand and explain localized histories by thinking about them as always connected to larger processes, hence as always and constitutively *global* (see Clarence-Smith 2015; Kalb and Tak 2006; Tsing 2000). Minescapes, in fact, provide peculiar examples of 'classic enclave economies that are, at one and the same moment, both deeply integrated into the global economy and also fragmented from the national space' (Bridge 2009; see also Arboleda 2020). This entails having in mind that broader analytical scales are necessary to understand how localized extractive histories are variously integrated into the broader global minescape, and how spatially dispersed histories and their diversified temporalities are synchronized in the broader global timescapes of extraction. Far from assuming the global perspective as a deterministic and mechanical explanation of localized geographies of extraction, we emphasize how variations in the spatial and temporal configuration of these geographies are intimately related to broader historical patterns of resource extraction and accumulation, natural depletion and socio-environmental degradation.

Connecting the Dots

This book originates from an event held at the University of Cagliari in November 2019.[13] Only some of the papers presented at the event were finalized as book chapters. The book includes scholars with different backgrounds, from anthropology to geography, from sociology to political science. Despite the 'global' ambitions of the book, the geographical reach of the case studies is limited to Africa (Democratic Republic of Congo), the Americas (Brazil, Bolivia, Peru, USA) and Europe (the Italian Alps, Arctic Norway and Spain), thus leaving outside the important mining regions of Asia and Australia. There is no apology for this geographical limitation, which we acknowledge as relevant. At the same time, we never thought of this limitation as something that would undermine the overall objective of the book. Indeed, we did not intend the 'global life of mines' as a comprehensive overview of mining and post-mining cases and typologies worldwide. Rather, our primary objective was to provide insights into resource extraction and its afterlives as concomitant, varying, coexisting and interrelated processes. The chapters offer a thought-provoking range of case studies to think about the multiple articulations of the timescapes of extraction within the global minescape. A common theme throughout these chapters is the insightful examination of the temporal dimension of mining and post-mining, their interconnectedness and lasting impact on global socioecological relations.

The chapters address contemporary issues in mining and post-mining regions; from the long-standing lethal consequences of uranium mining in US New Mexico (and beyond; Mesbah) to the contradictory entailments of mining-led urbanization in Kolwezi (Democratic Republic of Congo; Geenen) and the contested imaginaries of fossil modernities flowing through fuel smuggling practices between Peru and Bolivia (Ødegaard); from the state reimagination of coal mining regions in the Arctic Svalbard archipelago (Ødegaard) and the remining of old historical sites in Europe (Buu-Sao) to the grassroots and marginal pursuit of post-mining memories in Brazil (De Souza Santos); from contentious memories of mining in the 'touristified' Italian Alps (Zanini) to the erasure of Native history in the heritage conservation practices of abandoned mining sites in California's Death Valley (White).

Most of the cases either show the historical extent of extractive activities or illuminate their enduring effects. De Souza Santos details how the guided tours in the former goldmines in Ouro Preto (Brazil) recount stories of slavery, enacted through the spiritual manifestation of the dead enslaved mineworkers. These stories of dispossession are narrated by the racialized descendants of the enslaved inhabitants, who constituted around 80 per cent of the population during the eighteenth century, when mining peaked. Buu-Sao examines the case of the Riotinto mine in Andalusia, Spain, which is one of the oldest mining sites in the world. Extraction halted in the early 2000s, when the mine was declared a cultural heritage site, but resumed in 2015 as part of the ongoing European mining revival. Since then, extraction has coexisted with heritage processes, offering a distinct illustration of the entanglement of different mining temporalities. White demonstrates how boom and bust cycles in the making of the Western American frontier in the eighteenth century aligned with cycles of land dispossession and displacement of native populations. Mineral extraction was also a long-standing, centuries-old activity in the Alps, until its ultimate decline in the 1960s as part of the wider downturn of mining in Europe. The closures of European mines coincided with the mining boom in Africa, which attracted numerous Italian miners – as evidenced by Zanini in her chapter on Macugnaga. Geenen describes how the extractive activities in Kolwezi (Democratic Republic of the Congo), one of the main sites of cobalt extraction, are undermining the urban configuration that was precariously built on the mineral ore during the twentieth century. Cobalt is an essential component for the production of electric batteries, which has led to a significant increase in market demand. Uranium mining in New Mexico (USA) has a relatively brief history, beginning in the mid-twentieth century. However, the mining industry has created dramatic problems of radioactive contamination and environmental pollution. In 1979, a waste-filled earthen dam at the Church Rock uranium mine released more radiation than the Three Mile Island nuclear power plant accident in the same year. Mesbah details the networks of struggle and care that unfold in response to the long-lasting effects of radiation, reflecting on the temporal disjunctures of the timescapes of extraction, shaped by the logics of racial capitalism and the endurance of alternative socioecological visions.

In his perceptive and imaginative Afterword, David Kideckel comments on and expands the content of the book chapters in order to figure out new future research directions on mining and post-mining in our fevered, Anthropocenic world scenario. As he convincingly argues, there are at least three wide research areas in which anthropologists and other scholars might fruitfully engage with global mining and post-mining: the planetary, physical changes and the bodily human modifications related to mining and climate change; the sociopolitical implications associated with mining operations worldwide; and the growing significance of narrative and heritage in the mining sector (Kideckel, this book). These and other important aspects are consistent with and nicely integrate Jacka's identification of materiality, gender and remote mining as three significant areas of future mining research (Jacka 2018: 70).

In conclusion, with this book we intend to offer a framework and a number of significant examples of how theoretically and ethnographically to explore new and old mining and post-mining sites and operations and their uneven articulations (i.e. their social, cultural and political life) in a global context. The perspectives developed and the chapters assembled in this endeavour show how insightful it can be to bring mining and post-mining research into conversation, highlighting connections, disconnections and disjunctions. Consequently, we imagine the timescapes of extractions in wider spatial contexts, across a vast array of countries and continents, well beyond mining and post-mining as isolated sites and locations. Inspired by the seminal work of Eric Wolf (1997 [1982]), we propose to look at the world of mining and post-mining 'as a whole, a totality, a system, instead of as a sum of self-contained' mines and locations. Accordingly, we suggest reflecting on and exploring 'how this totality developed over time' and how each site and location is finally 'inextricably involved with other aggregates near and far, in weblike, netlike connections' at a global, planetary level (ibid., 385).

Acknowledgements

This book is the outcome of an event organized in the framework of the project 'Ecofrictions of the Anthropocene: Sustainability and Heritage-Making in Post-Industrial Reconversion', funded by the Italian Ministry of University and Research (Progetto di Ricerca di Interesse Nazionale [PRIN], Project number: 20155TYKCM). Antonio Maria Pusceddu received support from the Fundação para a Ciência e a Tecnologia (CEECIND/01894/2018/CP1533/CT0001) and the Centro em Rede de Investigação em Antropologia (UIDB/04038/2020).

Antonio Maria Pusceddu is senior researcher at the Centre for Anthropological Research (CRIA), University Institute of Lisbon, Portugal. He has conducted extensive field research in southern Europe. Labour, popular ecologies and social reproduction in industrialized regions are the focus of his current work. Recent publications include, 'Energopolitics of Transition: The Political Ecology of Anticipation in the Portuguese Lithium Rush' (*Anthropological Quarterly*, 2024); 'Southern Chronicles: The Political Ecology of Class in the Italian Industrial Periphery' (*Capitalism Nature Socialism*, 2022); 'Grassroots Ecologies of Value: Environmental Conflict and Social Reproduction in Southern Italy' (*Antipode*, 2020).

Filippo M. Zerilli is professor of anthropology at the University of Cagliari, Italy. Research interests include the history of anthropology, post-socialism, housing, social movements, pastoralism, humanitarianism and the anthropology of law and rights. He has conducted archival research in France, and field research in Romania, Sardinia and Kosovo. He is the author of *Il lato oscuro dell'etnologia* (CISU, 1998), editor of *Dalle 'Regole' al 'Suicidio'* (Argo, 2001), and co-editor of *European ethnology meetings* (ESI, 1998), and of *La ricerca antropologica in Romania* (ESI, 2003). Since 2015 he has served as editor-in-chief of *Anuac*, journal of the Italian society of cultural anthropology.

Notes

1. The engravings of the America series began to be published in 1590 by Theodore de Bry and his sons. The complete collection of plates was recently released by the German publisher Taschen (Van Groesen 2019). The engraving 'Washing gold' in Wolf's book corresponds to the first plate of the fifth volume of the series, which depicts the Spanish conquest of the Caribbean. The volume was published in 1595 and was based on *Historia del Mondo Nuovo* (1565), the account of the Milanese traveller Girolamo Benzoni. The original description of the plate was: 'African slaves are sent to the islands by the Spaniards to search for gold' (Van Groesen 2019: 230).
2. In this Introduction, we almost exclusively refer to mineral extraction. Likewise, nearly all chapters examine cases of mineral extraction (gold, copper, uranium, cobalt etc.), except for Ødegaard's chapter, which focuses on coal mining and fuel smuggling.
3. A good indication of this proliferation of studies on mining is the interdisciplinary journal *The Extractive Industries and Societies*, launched in 2014. For an exhaustive overview, in anthropology and beyond, see D'Angelo and Pijpers (2022a).
4 Calvão, Archer and Benya (2023a, 2023b) published two edited volumes that address separately the 'lives' and 'afterlives' of extraction.
5. For a brief and concise summary of the first three terms see, Haraway (2015); on the Wasteocene, see Armiero (2021); Plasticocene does not seem to have found a consistent theoretical elaboration yet, though it certainly proliferates in the world of arts and journalism.
6. Other minerals include copper, nickel, manganese, chromium, cobalt, zinc, rare earth, silicon etc.
7. For a critical discussion of extractivism (a system of accumulation that involves the transfer of wealth and raw materials from the periphery to the centre, associated with a colonial type of relation) and neo-extractivism (the assumption, in the periphery, that resource extraction can provide the basis for a nationally governed development), see Acosta (2013). It is worth noting that the term extractivism is also used to refer to a variety of domains unrelated to resource extraction *strictu sensu* (for example, Gómez-Barris 2017; Mezzadra and Neilson 2017).
8. It is widely assumed that mining activities have virtually disappeared from Europe in the past five decades, though this is not entirely true. The European mining industry still accounts for a significant supply of minerals but has become entirely dependent on import (specifically, from Asian countries) for today's strategic minerals – from lithium and graphite to rare earths. Other factors, from the massive mechanization of extraction and the considerable downsizing of the labour force, may account for

the marginalization (or even invisibilization) of a sector that accounted for much working-class conflict. A fact that may explain the 'political' demise of the mining industry and the mine specifically as a site of class struggle (see Buu-Sao, this book).

9. The wave of mine closures in Europe happened at the same time as mineral extraction was growing in Africa. This also affected labour mobility among European workforces. As shown by Zanini (this book), mineworkers from the Alpine mining region migrated to work in the Ugandan mining industry in the 1960s and 1970s. Literature about the African Copperbelt holds a special place in the anthropology of mining and urbanism, due to the work of a group of scholars later associated with the Manchester School (Ferguson 1999; Powdermaker 1962).
10. For example, the well-known Ok Tedi Mine project, 'one of the first conflicts between mining companies and communities to gain international prominence' (Kirsch 2014: 18).
11. In his detailed scrutiny of Marx's ecology, Paul Burkett (2014 [1999]: 112) points out how 'capitalism's accelerated throughput involves a conflict between the time nature requires to produce and absorb materials and energy versus the competitively enforced dynamic of maximum monetary accumulation in any given time period by all available material means'.
12. Wallerstein's world-system theory inspired more comprehensive elaborations of global environmental change, such as Moore's (2015) 'world-ecology' or Hornborg's (1998, 2012) 'ecological theory of unequal change' (see also Hornborg, McNeill and Martinez-Alier 2007).
13. The full list of participants at the event included: Paola Atzeni (University of Cagliari), Francesco Bachis (University of Cagliari), Alla Bolotova (University of Helsinki), Doris Buu-Sao (University of Lille), Kristien Geenen (Université de Liège), David A. Kideckel (Central Connecticut State University), Cristiano Lanzano (Nordic Africa Institute, Uppsala), Targol Mesbah (California Institute of Integral Studies), Cecilie Vindal Ødegaard (University of Bergen), Francesca Pugliese (Université de Liège/University of Leiden), Antonio Maria Pusceddu (CRIA/ISCTE-IUL), Tomasz Rakowski (University of Warsaw), Andreza A. de Souza Santos (University of Oxford), Felice Tiragallo (University of Cagliari), Pier Paolo Viazzo (University of Turin), Paul J. White (University of Nevada), Roberta Clara Zanini (University of Turin), Filippo M. Zerilli (University of Cagliari).

References

Acosta, Alberto. 2013. 'Extractivism and Neoextractivism: Two Sides of the Same Curse', in Miriam Land and Dunia Mocrani (eds), *Beyond Development: Alternative Visions from Latin America*. Amsterdam: Transnational Institute, pp. 61–86.

Adam, Barbara. 1998. *Timescapes of Modernity: The Environment and Invisible Hazard*. London: Routledge.

Arboleda, Martín. 2020. *Planetary Mine: Territories of Extraction under Late Capitalism*. London: Verso.

Armiero, Marco. 2021. *Wasteocene: Stories from the Global Dump*. Cambridge: Cambridge University Press.

Askland, Hedda H. 2018. 'A Dying Village: Mining and the Experiential Condition of Displacement', *The Extractive Industries and Society* 5(2): 230–36.

Ballard, Chris, and Glenn Banks. 2003. 'Resource Wars: The Anthropology of Mining', *Annual Review of Anthropology* 32: 287–313.

Barandiarán, Javiera. 2019. 'Lithium and Development Imaginaries in Chile, Argentina and Bolivia', *World Development* 113: 381–91.

Bridge, Gavin. 2004. 'Contested Terrain: Mining and the Environment', *Annual Review of Anthropology* 29: 205–59.

———. 2009. 'The Hole World: Spaces and Scales of Extraction', *New Geographies* 2: 43–48.

Burawoy, Michael et al. 2000. *Global Ethnography: Forces, Connections, and Imaginations in a Postmodern World*. Berkeley, CA: University of California Press.

Burkett, Paul. 2014 [1999]. *Marx and Nature: A Red and Green Perspective*. Chicago: Haymarket.

Burton, John. 1997. '*Terra Nugax* and the Discovery Paradigm: How Ok Tedi Was Shaped by the Way It Was Found and How the Rise of Political Process in the North Fly Took the Company by Surprise', in Glenn Banks and Chris Ballard (eds), *The Ok Tedi Settlement: Issues, Outcomes and Implications*. Canberra: National Centre for Development Studies, Pacific Policy Paper 27, and Resource Management in Asia-Pacific, RSPAS, Australian National University, pp. 27–55.

Calvâo, Filipe. 2011. 'When Boom Goes Bust: Ruins, Crisis and Security in Megaengineering Diamond Mining in Angola', in Stanley D. Brunn (ed.), *Engineering Earth*. Dordrecht: Springer, pp. 367–82.

Calvão, Filipe, Mathew Archer and Asanda Benya (eds). 2023a. *The Lives of Extraction: Identities, Communities and the Politics of Place*. Leiden: Brill.

———. 2023b. *The Afterlives of Extraction: Alternatives and Sustainable Futures*. Leiden: Brill.

Castillo, Gerardo, and David Breteton. 2018. 'The Country and the City: Mobility Dynamics in Mining Regions', *The Extractive Industries and Society* 5(2): 307–16.

Clarence-Smith, William G. (ed.). 2015. *Mining in Global History*, special issue of *Journal of Global History* 10(1): 1–76.

Conrad, Sebastian. 2016. *What Is Global History?* Princeton: Princeton University Press.

Crutzen, Paul J., and Eugene F. Stoermer. 2000. 'The "Anthropocene"', *Global Change Newsletter* 41: 17–18.

D'Angelo, Lorenzo, and Robert J. Pijpers. 2018a. 'Mining Temporalities: An Overview', *The Extractive Industries and Society* 5(2): 215–22.

D'Angelo, Lorenzo, and Robert J. Pijpers (eds). 2018b. Special Section: 'Mining Temporalities: Extractive Industries and the Politics of Time', *The Extractive Industries and Society* 5(2).

———. (eds). 2022a. *The Anthropology of Resource Extraction*. New York: Routledge.

———. 2022b. 'The Anthropology of Resource Extraction: An Introduction', in Lorenzo D'Angelo and Robert J. Pijpers (eds), *The Anthropology of Resource Extraction*. New York: Routledge, pp. 1–19.

Engwicht, Nina. 2018. '"It Can Lift Someone from Poverty": Imagined Futures in the Sierra Leonean Diamond Market', *The Extractive Industries and Society* 5(2): 260–66.

Ferguson, James. 1999. *Expectations of Modernity: Myths and Meanings of Urban Life on the Zambian Copperbelt*. Berkeley, CA: University of California Press.

Ferry, Elizabeth E., and Mandana E. Limbert (eds). 2008. *Timely Assets: The Politics of Resources and Their Temporalities*. Santa Fe: School for Advanced Research Press.

Gidwani, Vinay. 2012. 'Waste/Value', in Trevor J. Barnes, Jamie Peck and Eric Sheppard (eds), *The Wiley-Blackwell Companion to Economic Geography*. Oxford: Wiley-Blackwell, pp. 275–88.

———. 2013. 'Six Theses on Waste, Value, and Commons', *Social & Cultural Geography* 14(7): 773–83.

Godoy, Ricardo. 1985. 'Mining: Anthropological Perspectives', *Annual Review of Anthropology* 114(1): 199–217.

Gómez-Barris, Macarena. 2017. *The Extractive Zone: Social Ecologies and Decolonial Perspectives*. Durham, NC: Duke University Press.

Haraway, Donna. 2015. 'Anthropocene, Capitalocene, Plantationocene, Chthulucene: Making Kin', *Environmental Humanities* 6: 159–65.

High, Steven. 2021. 'Deindustrialization and its Consequences', in Michele Fazio, Christie Launius and Tim Strangleman (eds), *Routledge Handbook of Working-Class Studies*. London: Routledge, pp. 169–79.

Hornborg, Alf. 1998. 'Towards an Ecological Theory of Unequal Exchange: Articulating World System Theory and Ecological Economics', *Ecological Economics* 25(1): 127–36.

Hornborg, Alf. 2012. *Global Ecology and Unequal Exchange: Fetishism in a Zero-Sum World*. London: Routledge.

Hornborg, Alf, John R. McNeill and Joan Martinez-Alier (eds). 2007. *Rethinking Environmental History: World-System History and Global Environmental Change*. Lanham, MD: Altamira Press.

IEA. 2021. *The Role of Critical Minerals in Clean Energy Transitions*. World Energy Outlook Special Report. International Energy Agency. Retrieved 10 January 2023 from https://www.iea.org/reports/the-role-of-critical-minerals-in-clean-energy-transitions.

Jacka, Jerry K. 2018. 'The Anthropology of Mining: The Social and Environmental Impact of Resource Extraction in the Mineral Age', *Annual Review of Anthropology* 47(1): 61–77.

Kalb, Don, and Luisa Steur. 2015. 'Global South', in James D. Wright (ed.), *International Encyclopedia of the Social & Behavioral Sciences*, vol. 10, 2nd ed. Oxford: Elsevier, pp. 186–91.

Kalb, Don, and Herman Tak (eds). 2006. *Critical Junctions: Anthropology and History beyond the Cultural Turn*. New York: Berghahn.

Kideckel, David A. 2018. 'Coal Power: Class, Fetishism, Memory, and Disjuncture in Romania's Jiu Valley and Appalachian West Virginia', *Anuac* 7(1): 67–88.

Kirsch, Stuart. 2010. 'Sustainable Mining', *Dialectical Anthropology* 34(1): 87–93.

———. 2014. *Mining Capitalism: The Relationship Between Corporations and Their Critics*. Berkeley, CA: University of California Press.

Knapp, Arthur B., Vincent C. Pigott and Eugenia W. Herbert (eds). 1998. *Social Approaches to an Industrial Past: The Archaeology and Anthropology of Mining*. London: Routledge.

Komlosy, Andrea. 2018. *Work: The Last 1000 Years*. London: Verso.

Labban, Mazen. 2014. 'Deterritorializing Extraction: Bioaccumulation and the Planetary Mine', *Annals of the Association of American Geographers* 104(3): 560–76.

Lanzano, Cristiano. 2022. 'Sustainability', in Lorenzo D'Angelo and Robert J. Pijpers (eds), *The Anthropology of Resource Extraction*. New York: Routledge, pp. 149–66.

Limbert, Mandana. 2016. 'Liquid Oman: Oil, Water, and Causality in Southern Arabia', *The Journal of the Royal Anthropological Institute* 22(1): 147–62.

Luning, Sabine. 2012. 'Corporate Social Responsibility (CSR) for Exploration: Consultants, Companies and Communities in Processes of Engagements', *Resources Policy* 37(2): 205–11.

———. 2018. 'Mining Temporalities: Future Perspectives', *The Extractive Industries and Society* 5(2): 281–86.

Machado Aráoz, Horacio. 2018 [2014]. *Potosí, el origen: Genealogía de la minería contemporánea*. Buenos Aires: Editorial Maldulce.

Metcalfe, A.W. 1990. 'The Demonology of Class: The Iconography of the Coalminer and the Symbolic Construction of Political Boundaries', *Critique of Anthropology* 10(1): 39–63.

Mezzadra, Sandro, and Brett Neilson. 2017. 'On the Multiple Frontiers of Extraction: Excavating Contemporary Capitalism', *Cultural Studies* 31(2–3): 185–204.

Mitchell, Timothy. 2013. *Carbon Democracy: Political Power in the Age of Oil*. London: Verso.

Moore, Jason W. 2007. 'Silver, Ecology, and the Origins of the Modern World', in Alf Hornborg, John R. McNeill and Joan Martinez-Alier (eds), *Rethinking Environmental History: World-System History and Global Environmental Change*. Lanham, MD: Altamira Press, pp. 123–42.

———. 2015. *Capitalism in the Web of Life: Ecology and the Accumulation of Capital*. London: Verso.

Nixon, R. 2011. *Slow Violence and the Environmentalism of the Poor*. Cambridge, MA: Harvard University Press.

Oakley, Peter. 2018. 'After Mining: Contrived Dereliction, Dualistic Time and the Moment of Rupture in the Presentation of Mining Heritage', *The Extractive Industries and Society* 5(2): 274–80.

Patel, Raj, and Jason W. Moore. 2018. *A History of the World in Seven Cheap Things*. London: Versus.

Powdermaker, Hortense. 1962. *Copper Town: Changing Africa, the Human Situation on the Rhodesian Copperbelt*. New York: Harper Colophon.

Revette, Anna C. 2016. 'This Time it's Different: Lithium Extraction, Cultural Politics and Development in Bolivia', *Third World Quarterly* 38(12): 149–68.

Richardson, Tanya, and Gisa Weszkalnys. 2014. 'Introduction: Resource Materialities', *Anthropological Quarterly* 87(1): 5–30.

Rubbers, Benjamin (ed.). 2021. *Inside Capitalism: The Micropolitics of Work on the Congolese and Zambian Copperbelts*. Woodbridge: Boydell & Brewer.

Strangleman, Tim, James Rhodes and Sherry Linkon. 2013. 'Introduction to Crumbling Cultures, Deindustrialization, Class, and Memory', *International Labor and Working-Class History* 84: 7–22.

Thompson, Edward P. 1963. *The Making of the English Working Class*. London: Penguin.

Tomba, Massimiliano. 2009. 'Historical Temporalities of Capital: An Anti-Historicist Perspective', *Historical Materialism* 17: 44–65.

Tsing, Anna. 2000. 'The Global Situation', *Cultural Anthropology* 15(3): 327–60.

———. 2005. *Frictions: An Ethnography of Global Connection*. Princeton, NJ: Princeton University Press.

Vaccaro, Ismael, Krista Harper and Seth Murray (eds). 2017. *The Anthropology of Postindustrialism: Ethnographies of Disconnection*. London: Routledge.

Van der Linden, M. 2008. *Workers of the World: Essays Toward a Global Labor History*. Leiden: Brill.

———. 2012. 'The Promise and Challenges of Global Labor History', *International Labor and Working-Class History* 82: 55–76.

Van Groesen, Michiel (ed.). 2019. *Theodore de Bry. America: The Complete Plates 1590–1602*. Cologne: Taschen.

Wallerstein, Immanuel. 1974. *The Modern World-System, vol. I: Capitalist Agriculture and the Origins of the European World-Economy in the Sixteenth Century*. New York and London: Academic Press.

Wolf, Eric R. 1997 [1982]. *Europe and the People Without History*. Berkeley, CA: University of California Press.

World Bank. 2020. *Minerals for Climate Action: The Mineral Intensity of the Clean Energy Transition*. Washington. Retrieved 10 January 2023 from https://pubdocs.worldbank.org/en/961711588875536384/Minerals-for-Climate-Action-The-Mineral-Intensity-of-the-Clean-Energy-Transition.pdf.

— Chapter 1 —

Frontier Spaces in the Arctic and the Andes

The Miner, the Smuggler and Performances of (Post-)Extractivism

Cecilie Vindal Ødegaard

While the current frontiers of resource extraction appear as constantly expanding into new terrains and areas of life, their movements are increasingly accompanied by policies of environmentalist preservation, heritagization and climate mitigation. A central claim in this edited volume is that the frontiers of resource extraction cannot be seen as isolated from environmentalist narratives and policies but as interrelated – and as producing highly diversified socioecological geographies of territory and labour. Indeed, the policies of extractivism as well as 'post-extractivism'[1] are both framed within narratives of development and progress and facilitate the (re)productions of cosmopolitan centres and global frontiers, including the differentiations of spaces and identities (Franquesa 2018; Tsing 2005). This chapter explores how (post-)extractivist policies and narratives partake in staging society and territories – and hence also social identities – in particular ways; by bringing changes to public spaces, infrastructures and workplaces, and celebrating (or externalizing) particular kinds of identities through ways of mobilizing, organizing and/or dislocating labour. Being concerned with the uneven geographies of (post-) extractivism, the chapter examines the production of space and identity in two different contexts: the Arctic (the Svalbard archipelago, Norway) and the Andes (the Peruvian-Bolivian border areas). I draw on my ethnographic fieldwork[2] in both these contexts to discuss what we can learn by exploring their commonalities and differences; that is, as cases of (post-)extractivist frontiers that entail complex coexistences of practices attempting to define, make claims to and re-create spatial domains: what characterizes the interconnections between the production of space, identities and temporalities at global frontiers? What are the differentiated personae and identities narrated through the – often contested – claims to frontier spaces? The chapter discusses the emergence of two different and somewhat anachronistic personae; the former miner in Svalbard and the smuggler (*contrabandista*) of fuel in Peru. In the global geographies of (post-)extractivism, these personae are ambiguously made part of the dominant narratives of progress and development. The chapter suggests that the emergence of such anachronistic personae represents an important intake for ethnographic explorations of the connections and disconnections between (post-)extractivist policies and narratives.[3] The chapter discusses how both the former miner and the small-scale smuggler in the Andes are made into central personae in the (re)makings of (post-)extractivist spaces; that is, one domain that is re-created as a post-extractivist and environmentalist showcase (Svalbard), and another domain that is contested in its supposed marginality to the extractivist state[4] (that is, the Peru-Bolivia border areas).

The making of identities through (post-)extractive narratives and performances entails temporal as well as spatial dimensions, as extraction is accompanied by particular narratives about time and the future in ways that locate people in time as well as space. This is evident not only in the narratives of early modern industrial eras celebrating the hard work and bravery of mine workers as key to national progress and development, but perhaps even more evident in imaginings of the people placed outside, excluded from or dispossessed by resource frontiers, such as (in many cases) Indigenous peoples – or laid-off workers in post-mining sites. Importantly, Ferry and Limbert (2008) argue that the acts of making (and unmaking) a resource produce certain temporal effects, as the products and values created through such ideational systems (or resource imaginations) also frame the past, present and future in particular ways. (Post-) extractivist spaces themselves are thus defined and characterized by the ebb and flow of the 'timely assets' that are extracted there (Ferry and Limbert 2008; see also Franquesa 2018). Meanwhile, current policies and narratives of post-extractivism, or energy transition, do appear to entail a radical shift of industrialism's narrative framing. While the idea of industrial progress, emerging in the late 1700s and beginning of the 1800s, is a notion of linear development and growth, the current emphasis on energy transition (though differentiated and differentiating) can be considered a prefigurative discourse, anticipating change (including loss and decay) while seeking to prefigure what such changes might look like and how they might be dealt with. Elsewhere (Ødegaard 2022), I have therefore suggested that attention to the narrative and performative dimensions of proclaimed transition can facilitate our understanding of how the prefigurative transition framework is reshaping both temporal and spatial dimensions of the industrialization narrative.

Spatialization can take place through various ways of seeing, knowing and governing – and may hence produce particular identities, subjectivities and personae, through the production of spatial boundaries and borders, the displacement or erasure of people and livelihoods (Meyer 1996), the fixation of people in particular spaces (Whitesell 1996), or by holding inhabitants to particular discursive standards (Hilgers 2010; Igoe 2005; Rose 2007), as I will return to. While the production of space takes place at the crossroads between top-down and bottom-up processes, the hegemonic dimensions of space are defined by histories of extractivism, (settler) colonialism and increasingly through policies of environmentalist preservation and/or climate mitigation (Büscher and Fletcher 2020). These hegemonic forms tend to (re)produce particular spaces as 'frontiers', envisioned as 'empty places' ready for resource exploitation or as sites of 'pure, pristine wilderness' – in ways that make them easier to govern, exploit (Tsing 2005; West, Igoe and Brockington 2006) or reimagine. In this regard, Tsing (2003) has importantly argued that the notion of 'resource frontiers', as it is generally used, can serve to legitimize processes of 'freeing' land for extraction and dispossession. Similarly, and recognizing that the notion of 'frontier' may implicate a linear movement across space, Rasmussen and Lund (2018) propose the term 'frontier spaces' to pinpoint a continual emphasis on the discovery or invention of new resources at global frontiers. Indeed, the notion 'frontier spaces' is useful in addressing frontiers as historically constituted spaces that are continually being remade, reclaimed and/or reinvented for purposes of value extraction. This may include also more intangible values, such as the extraction from the very lives, bodies, homes and environments of local populations (Ødegaard and Rivera Andía 2019). In this regard, the spatial domains addressed in this chapter are constituted rather ambiguously as the frontiers of the extractivist state; that is, one domain which is redefined as a post-mining site and the other as a territory that defies and evades the state's accumulative structures.

In recent decades, studies of neoliberalism have been influenced by the governmentality approach, drawing on Foucault and focusing on the techniques, technologies and discourses of neoliberal governance and its production of particular kinds of (ideal) subjects (Hilgers 2010; Rose 2007) that 'fit' or support the dominant narrative. While one should not overestimate the extent to which such articulation of 'ideal subjects' results in the actual creation of these selves (Winkler-Reid 2017), the powerful effects of dominant narratives and policy measures on subjective identities, desires and affects are perhaps especially evident in their externalizing effects; that is, in the case of people who somehow 'fall outside' or are further externalized[5] by such measures, as will be my emphasis in this chapter.

In what follows, I discuss the case from Svalbard, with a focus on the making of spatial identities in post-extractivist narratives about the archipelago. This case raises questions regarding the management of the *remains of* a particular form of extractivism, namely coal mining. Coal mining sites on Svalbard have been gradually closed down, although this may well be to accommodate for other forms of extractivism; for instance, the commercialization of nature and mining history through the tourist industry (Sokolickova and Eriksen 2022). Next, I somewhat briefly[6] discuss my case from Peru, emphasizing the production of space and identities in the contested circulation of fuel at the borders with Bolivia, followed by some comparative reflections on how narratives of (post-)extractivism (re)create differentiated frontier spaces and identities.

Post-Extractivist Narratives in Svalbard

Most settlements in Svalbard were built around and defined through mining activity – in the middle of the Arctic wilderness; for example, like the Norwegian settlements Longyearbyen, Svea, Ny-Ålesund, and Russian Barentsburg and Pyramiden. Miners in Svalbard and beyond have often been represented as the hard-working heroes of industrial modernization, having provided the raw materials of energy and wealth through their mastery of moving and working below the surface of the earth (Fløgstad 2007). Now, in the current context of environmental instabilities, climate change and uncertain futures for the fossil fuel markets, coal mining sites in many parts of the world have closed down – and in some cases even been dismantled, like in Svea, Svalbard. While some authors have noted how the call for a transition to less carbon-intensive systems sometimes casts the people – whose work brings fossil fuels into being – as somehow immoral (High and Smith 2019: 19), the mineworker in Svalbard is rather made to represent the negative temporal other of the new narrative of post-extraction; whose[7] labour efforts are celebrated, while their place in time are being redefined.

In 2017, the Norwegian government announced that the coal mines in Svea would close and later that all the mining infrastructure should be dismantled, including the Svea settlement itself. The stated goal was to remove the traces of human activity and restore the place to its original, 'natural' state, with the intention that the area should appear as uninfluenced by humans as possible, except, that is, for a few older buildings and roads built before 1946, as all structures from before 1946 are protected as cultural heritage in Svalbard. When explaining the decision, the Minister of Trade and Industry at the time, Monica Mæland, stressed the economic motive behind the decision: the coal market was down lower than expected, and a continuation of the mines could therefore not be justified. Further, the activities connected to the clean-up (and 'returning to nature') would give the community time to adjust to the transition from coal mining to other economic activities and energy sources. As a result, the mining community in Svea, located to the south of Longyearbyen, has now been dismantled. The dismantling started in 2018, with around fifty people working at the site to remove the buildings and infrastructure. They worked 14 days on and 14 days off, as was usual in the times when the Svea mines were operating. The work of dismantling in many ways followed the temporal rhythms of mining work itself – only the purpose and nature of the labour was different. Most of the workers who undertook the dismantling were former miners, many of whom used to mine in Svea and were known as 'the workers who stayed behind'. This phrasing has a double edge to it, as these workers were not only literally staying behind in Svea but were working to dismantle their previous workplace community. As I have argued elsewhere (Ødegaard 2022), the dismantling and 'turning back to nature' in Svea is part of an environmentalist narrative about Svalbard: a narrative about the archipelago as a showcase for environmentally friendly initiatives 'after mining'.

This is the first time that an entire mining community in Svalbard (or Norway) has been dismantled and 'returned to nature'.[8] Other abandoned mining settlements in the archipelago have been maintained: for example, the mining settlement in Ny-Ålesund now serves as a research station, and one of the old mines outside Longyearbyen, Gruve 3, is open for guided tours. Guided tours to the Russian mining town of Pyramiden are also available; the town was left to decay after abrupt abandonment in 1998, transforming it into what Fløgstad describes as 'a stereotype of yesterday's utopian thinking, now solidified into pure building mass' (2007: 56). The Svea community, on the other hand, has been dismantled and removed – with some of the materials and structures sent for re-use in Longyearbyen or for international resale – and the community resurrected through a 3D reconstruction based on extensive photo materials in an attempt to digitally re-create as much as possible everyday life as it played out in Svea.

The dismantling and 'returning' in Svea is a spectacle of coal mining's termination in the archipelago, a social drama of our Anthropogenic times (see Ødegaard 2022) and part of the remaking of Svalbard as a site of environmentalist initiatives and solutions. Indeed, the 'return' in Svea has taken place in a context where environmental instabilities and climate change are high on the agenda, especially after the lethal avalanche in Longyearbyen in 2015. The situation of increasingly unstable weather and sea-ice conditions simultaneously opens up geopolitical concerns and questions about sovereignty, presence and ownership in the Arctic, with new opportunities related to shipping, mineral extraction and other resource exploitation. New questions are also being raised about the foundation for settlements under challenging Arctic conditions; for example, energy supply and Svalbard's long-standing reliance on coal. Several changes are already underway: Gruve 7, close to Longyearbyen, is the only Norwegian mine still operating in Svalbard in addition to the mining activities in Russian Barentsburg. The mining in Gruve 7 is soon to end,[9] and the coal-power plant in Longyearbyen has been phased out, with the intention of replacing it with a more climate-friendly solution.[10] The power plant has been run by diesel since autumn 2023, as a temporary solution while awaiting a greener energy source alternative. However, much is still uncertain concerning security of supply and price levels for consumers – and regarding the new energy alternative itself because the technology they need is simply not yet 'mature', as one of the SN administrators commented to me. Tests, reports and meetings seem to be leading primarily to more uncertainty – for now. Meanwhile, these attempts to initiate energy transition in Svalbard create new frames for narratives about civilization's 'frontiers'. This way of (re)making of Svalbard as a spatial frontier differs from the frontiers of the Andes discussed below, in that the latter is a contested space of upholding official master narratives of extractivism, while Svalbard is made into a test case for post-mining transition and innovation.

The project in Svea was presented as Norway's most comprehensive environmental clean-up project to date, at an estimated cost of 1.9 billion NOK. The government's goal of bringing the place back to its 'original, natural' state was accompanied by the inclusion of the area into the Nordenskjold Land National Park, re-named in 2021 as Van Mijenfjord National Park. The Norwegian company Store Norske Spitsbergen Kullkompani (here referred to as Store Norske or SN) was responsible for the dismantling and restoration in Svea, finalized early autumn 2023. Up to 350 persons lived and worked there (most miners), so the decision removed a whole society – with its houses, furniture, tractors and the like – in addition to mining infrastructure, and included the cleaning-up of various forms of waste material surrounding the community and the mines. In this and other ways, Svea was quite literally unmade as a site of coal extraction and thus made part of the unmaking of carbon reliance – and Svalbard is hence being reinvented as a space for post-extractivist transition. Another element of the environmentalist narrative about Svalbard was the opening of the Seed Vault in 2008. The Seed Vault is presented along the lines of being Norway's and Svalbard's 'gift to the world', providing safe storage for seeds from all over the world and hence taking care of humanity's food supply in an uncertain future. The so-called 'Doomsday vault' soon got something of a superstar status, and then prime minister Jens Stoltenberg compared it to innovations like Tesla and Bitcoin and referred to it as 'a modern Noah's Ark'. Interestingly, and although only special invitees can enter the vault, it soon became nearly like a modern pilgrimage site, and local guides say that tourists sometimes come just to gaze at the exterior of the vault and even camp there for days without much around than the vault itself. Indeed, considering the narrative and performative dimensions of transition, these and other initiatives can be taken to indicate how 'transition' entails a lot more than 'just' transition to a new energy source. It indicates how post-extractivist narratives stage society, place and time in particular ways.

From Resource Frontier to Nature's Frontier

In contrast to other Arctic areas – and in contrast to the Andean region discussed below – there is no Indigenous population in Svalbard. This and the Svalbard Treaty's significance for political decision-making in the archipelago make questions of inhabitancy, presence and entitlement play out differently than in regions where Indigenous people's historical and existential connection to place entails a particular vulnerability and at the same time specific rights (at least in theory). Since the catch expeditions in the late 1600s, human presence in Svalbard has comprised of 'Arctic scrambles' (Dodds and Nuttall 2016), a term pinpointing how the region has been constituted through the search for resources and scientific exploitation. In Svalbard, and partly due to the specificities of the Svalbard Treaty, these 'scrambles' have produced continual strategies of colonization also through the promotion of particular kinds of inhabitants: first for purposes of resource extraction, and more recently for purposes of national presence and innovation, as I return to.

The Svea field was established in 1917 by the Swedes, following the discovery of coal in the early 1900s. In 1934 it was bought by Store Norske, two years after the Soviet Union bought Barentsburg. Coal production in Svea was intensified with the initiation of a new mine, Svea Nord, in 2001. The establishment of this mine carried particular significance for long-term inhabitants in Svalbard because it represented an important Norwegian commitment to coal mining in the archipelago and further job opportunities. From then on, Svea was the major coal production site in Svalbard. In 2014 another new mine was finalized in Svea by Store Norske, the Lunckefjell mine, with significant investment (1.2 billion NOK).[11] This mine, like the Svea Nord mine, came with promises of work opportunities and growth in Svalbard. Soon after, however, came the fall in oil prices (affecting also coal prices), and production was put on temporary hold. In 2017, the Norwegians decided to stop production in Lunckefjell and Svea Nord for good, despite recent investment. In many ways, the dismantling in Svea was a pivotal point for Svalbard's transition from a mining community to a showcase for the future, marking the end of one era and the beginning of another. In this regard, the work of cleaning-up and 'returning' in Svea can be considered a form of 'temporal labour' in that it aimed to create a particular temporality to the place through the work to preserve the area for the future by bringing it 'back to what it once was'. The 'returning' quite spectacularly marked coal mining as a remnant of history, making the miner a historical figure and bringing into reconsideration ideas and expectations related to Svalbard as a mining society, mining as highly valued labour and the miners' central role in the archipelago. Mining work itself, as lived life, has been remade as memory, and as I have illustrated elsewhere (2022), nature in the area has been at one and the same time remade as an object of human design – and re-created as 'wild, pristine nature'.

Signed in 1920, the Svalbard Treaty established the sovereignty of Norway over the archipelago and gave the country particular rights and responsibilities in Svalbard as a territory under Norwegian jurisdiction, but with a principle of equal treatment that meant all signature countries had equal rights to live and entertain commercial interests there. In principle then, all signature countries can exploit the natural resources. Since the signing of the Treaty, human activity in Svalbard has centred around coal mining, where Norway and Russia long had a bilateral relationship (Totland 2016: 46). While at first coal mining was a goal in itself, it later became part of a strategic demarcation of national presence and visibility: first due to the archipelago's significance as an Eastern/Western outpost during the Cold War, and later related to the territory's increased geopolitical significance.

Especially since the 1990s, environmental regulation in the archipelago has become increasingly important not only for the protection of the natural environment but also as a means of governance (Saville 2019) and the demarcation of a 'Norwegian presence' (Ødegaard 2022). The Svalbard Environmental Law was introduced in 2001 as a commitment to a central principle in the Svalbard Treaty: Norway has a particular responsibility in the protection of the natural environment in the archipelago. This environmental responsibility allows Norway to mark its national presence by other means than its historically established inhabitation, settlement and extractive labour: that is, it may mark its presence by human absence (Ødegaard 2022) and keeping other Treaty-nations out of the area through expansion of the national park.

Not surprisingly, the announcement that the Svea mines would be closed and dismantled provoked strong reactions from the miners in Svalbard. Some even compared their own reactions to the heartbreak after a love affair: first shock, followed by gradual habituation for many. Miners described the loss not just of a workplace and job opportunities but also of a place of memories and lived connections (see also Ødegaard 2022). For them, the dismantling turned Svea from a promise of work solidarity and income into a remnant of the past. While well aware that there is a particular 'life cycle' to all mines, most miners would have preferred the Svea mines to have been emptied before being closed. There were also uncertainties about the future. One of Store Norske's coordinators in Svea told me that the strong reactions among miners were to be expected given uncertainties regarding current and future jobs: 'their work positions have been redefined. Even if their salaries have not been reduced, they receive less dirt surcharge than they did as miners.' Nobody was to lose their jobs, SN announced, but for some it was not clear what this would entail, especially for the older miners. Several people I spoke with (both miners and others) claimed that there was a particular sense of work unity (*samhold*) in Svea due to its challenging and often dangerous nature. Miners from Svea got to know the area and the landscape in a different way than most people, as they worked and lived in this inaccessible area during all seasons and often for years, hence knowing how the landscape and the climate changed during the year. However, it appears that miners were not consulted regarding the 'returning' project, and some people I spoke with raised questions about what would happen to this knowledge of the area once Svea was removed. One previous miner commented that it seemed to be ending up in digital archives and 'turned into something else'. After the dismantling, a contact at SN also said that even she, who used to know the Svea community and workplace very well, no longer recognized the place – so others would certainly have difficulties understanding that there used to be a whole, well-functioning community there.

The dismantling proclaimed a shift in Svalbard, a shift towards more environmentally friendly energy solutions while breaking with established arrangements and defined relations between different actors and parties. Coal mining is gradually being replaced by activities related to tourism, research and education, which have been expanding especially since the 1990s. Store Norske is redefining their role in the archipelago, from mining to the management mainly of properties and logistics. The company is also trying out new, environmentally friendly solutions, such as the trial project at Isfjord Radio and the testing of wind and solar power as well as thermal heat storage. Store Norske is central also to the management of mining history in Svalbard, maintaining abandoned mining structures and organizing guided tours of Gruve 3. In the wake of the dismantling, questions were being raised about how Norwegian authorities intend to mark 'national presence' now that the mines have closed down.

Longyearbyen is and has been characterized by a high population turnover. There is great variation in how long people remain: some for shorter periods in relation to specific work opportunities, others for several years, and some with family ties to the archipelago that go far back. Longyearbyen has also been developing into a highly international location, especially since the 1990s, and is now home to people from many regions of the world. Access to housing in Longyearbyen represents a challenge, however, especially for those employed in the private sector, which includes many of the non-Norwegian citizens. From being a company town, Longyearbyen is now developing other forms of inequality, separation and othering. Meanwhile, there is a common term used when referring to people living in Svalbard; *Svalbardianere* (or 'Svalbardians'). Among both short- and long-term Svalbardians, there is often strong engagement with questions of community development, change and environmental management in the archipelago.

Indeed, miners are not the only inhabitants critical of the closure and dismantling in Svea. Some opposed the closure of production from the start, and others wanted at least the maintenance of community structures in Svea. Many raised questions about the consequences that the closure in Svea would have for Longyearbyen, as people feared that closely-knit social bonds would be weakened further by a potential rise in the already high population turnover. The ongoing expansion of environmental regulation in Svalbard (as well as the inclusion of the Svea area in the national park) has caused further concerns for increased limitation of people's movements within Svalbard's grandiose landscape. Some say they get the impression that the need to 'build community' in Svalbard is not being recognized and that Norwegian authorities are trying to remake the archipelago into some kind of 'climate model' at the expense of inhabitants' need for closely-knit relations and possibilities to use the landscape. In fact, a widespread impression among inhabitants is that the official narrative about the clean-up in Svea is a way to 'greenwash' or conceal other, more strategic political motives; that is, it is another way of marking the 'Norwegian presence' and keeping other nations out. Indeed, many fear the societal implications of recent government decisions, raising questions about the direction of Norwegian Svalbard politics.

The current emphasis on nature and environment in governance and regulation contributes in many ways to the remaking of Svalbard from 'resource frontier' to 'nature's frontier'. This way of managing the environment through regulation and rewilding is certainly not politically or ideologically 'neutral', as has similarly been demonstrated in literature on national parks (Tsing 2005; West, Igoe and Brockington 2006), but it can be considered a techno-administrative fix that produces particular kinds of subjects. For instance, Tsing (2005) comments on how the protection of nature in areas apparently peripheral to political centres produces particular centre versus periphery mechanisms: in the so-called global 'frontiers', the preservation of nature can contribute to maintaining an image of political centres as cosmopolitan. Political centres, she argues, are hence often produced in contrast to the 'local people' – who are objectified as subjects of scientific inquiry as well as forms of governance considered appropriate to peripheral places or areas that appear to be the frontiers of civilization (2005: 141). In Svalbard, as 'nature' is made into a central entity both in the demarcation of national presence and in narratives about Svalbard as an environmental showcase, certain expectations are produced of the Svalbardian as a particular kind of person: not an industrial worker or miner but preferably an environmental expert, governable but innovative and preferably with ties to Norway's mainland.

Creating Innovative Subjects – and Externalizing the Miner

The narrative of extractivism in Svalbard envisioned miners as the heroes of industrialism and valued for their hard, manual work. In an interview, one of the previous miners in Svea described how they used to wash and shower at Lompen (now a shopping centre) after returning from a shift, exclaiming: 'Then we turned back into humans' (my translation). Miners told me that it was especially the miners who worked 'inside' the mines that enjoyed particular prestige, not just because this work was considered particularly demanding but also because they used to be paid more than those who worked 'outside'. When the miners finished their shift and returned to Longyearbyen for their free time, there used to be a good, festive atmosphere in town. There is still a sense of Longyearbyen as a 'last frontier', and although mines are being closed, the aesthetics of the somewhat rough company town are actively drawn upon by various, very active leisure time organizations. Concerts are sometimes arranged in Gruve 7, the still active mine outside Longyearbyen. At one concert I attended, a local choir wore miners' clothes and helmets and in various ways made reference to Longyearbyen as a company town. In this and other ways, company town aesthetics are often drawn upon to *stage* Longyearbyen as a community. Various ethnographers have demonstrated how the intensification of uncertainty in post-Fordist as well as postsocialist societies often has a temporal character and grounding in a nostalgia for the past and 'the standard employment relationship' ideal (Muehlebach 2011, and see, for example, Cowie and Heathcott 2003). Pinpointing such orientation as the 'post-Fordist affect', Muehlebach (2011) describes how ex-factory workers in Northern Italy engage in voluntary labour as a means of coping with the loss of jobs and the resulting lack of self-esteem and identity, hence re-enacting a form of social belonging. In Longyearbyen, there is an engagement in leisure organizations and miners' aesthetics that similarly re-enacts a form of social belonging; not necessarily or primarily with reference to 'the standard employment relationship' but by enacting belonging to Longyearbyen and Svalbard as a place.

The decision to close the Svea mines soon spurred debates about 'who belongs' there. A widespread sentiment has been that the miners *are* Svalbard, or that they are the core of the Svalbard settlements and central for maintaining the continuity of social bonds and mutual help, qualities that many see as characterizing the place. Others emphasize that the tourist guides and researchers are the new 'Svalbardians' and suggest that a high turnover is a good thing because new people with new ideas is necessary to ensure innovation. Alongside environmental regulation, Norwegian authorities have, especially since the 1990s, promoted a family-oriented community in Longyearbyen, accommodating for kindergartens and after-school activities and the like. Still, Longyearbyen is not a life course society but a 'working community'.[12] There is very much a direct governance of the archipelago through the local governor (Sysselmesteren) and Store Norske in particular. With the current emphasis on energy transition and climate change adaptation, this governance is increasingly directed towards the promotion of innovation. So-called 'Norwegian presence' then is increasingly secured through the accommodation of the creative, innovative subject. Indeed, the current emphasis on innovation, adaptation and resilience can be seen to reinforce this emphasis on a particular kind of subject. To put it somewhat crudely, with current narratives about Svalbard as an environmental showcase, there seems to be a need also to 'innovate' its inhabitants. It appears to take the form of another kind of settler colonialism, not in the form of lawlessness and the displacement of Indigenous populations but by a recolonization of the place through environmental management – and the accommodation for particular kinds of inhabitants. This version of settler colonialism is characterized by increased regulation – of nature management, commodity flow and access to democracy – and a strong state ownership and presence, tendencies that these days are being intensified vis-à-vis a Russian presence due to Russia's war on Ukraine.

While the aim of the digitalization project in Svea was to preserve the memory of the mining community by creating a virtual reconstruction of the place, it proved more difficult than originally envisioned to decide what this should entail. It was not certain, for instance, if the project should result in a digital exhibition or a game about mining, and consequently the project was delayed. During the dismantling process, an automatic camera was placed in Svea, taking hundreds of pictures per hour, in order to document the dismantling process and make sure 'nothing was lost', as Store Norske formulated it. The reason, according to one of my contacts, was partly that SN wanted to document not only the period before 1946, as the industrial revolution in Svea actually took place in the 2000s, and they wanted to avoid an impression that mining in Svea had ended in 1946. As I was talking with one of the previous miners in Svea, he recognized this wish to secure the memory of the mining community but also pointed out that the accumulated knowledge about the mining work itself would nonetheless disappear with the individual miners. One is tempted to ask what all this documentation is for – and if more emphasis has been placed on preserving the tangible (and, in the case of Svea, the visual) traces of mining than on intangible heritage (see, for example, Gerlach and Kinossian 2016), such as the knowledge, narratives and experiences of the miners themselves. As the dismantling and digital heritage making in Svea has brought the significance of mining work and the miners' central role in the archipelago into reconsideration, mining work as lived life has been remade as memory. At the same time, mining work is made subject to representations of particular workers' aesthetics through an emphasis on the physically hard and dangerous work forms. This takes place through the preservation of tangible heritage – and the tourist industry's focus on recapturing the 'memory of mining', such as through guided tours of Gruve 3 and of other industrial heritage sites – as well as the organization of particular workout sessions in the gym that replicate the physicality of miners' work. While such arrangements may contribute to remembering the labour and everyday lives of miners, they may reduce heritage and memory to objects for commercialization. Moreover, by constituting part of a multifaceted performance of a post-mining narrative about Svalbard, these arrangements contribute to externalizing mining and miners by making coal mining a remnant of history and the miner an exotified figure of the past. The figure of the miner is thus turned into the anachronistic 'other' but at the same time made central for representations of Svalbard through heritagization and tourist ventures. In assuming or creating this alterity, the institutionalized heritagization of industrial history creates the foundation for new innovative solutions and initiatives.

While the dismantling was sorrowful for many, as noted, others saw it as an opportunity. Mari and her colleagues at SN tried to turn things around from feelings of nostalgia to feelings of possibility, as she formulated it. This was facilitated by an emphasis on the re-use and re-sale of the materials and equipment from Svea, for which Mari was in charge. The workers involved in the dismantling and returning in Svea had to take a 'green-card' and 'brown-card' course to learn how to work and mould the terrain without damaging the landscape or protected buildings. This way the workers took more ownership of the process, according to SN representatives. These and other initiatives illustrate how different actors in Longyearbyen recognized the need to try and compensate for the experience of loss and turn the dismantling into something positive – in emphasizing environmental solutions and training workers. 'You've got to start somewhere', and 'this is a way to learn', as one of my contacts said in response to the critique that the project in Svea was just 'symbolic politics' (*symbolpolitikk*). At the same time, these initiatives have contributed to externalizing the persona of the miner. Indeed, as the narratives of extractivist industrialism are being retold in the context of climate change and environmental instability, the figure of the miner is unmade as a central representative of the nation's growth and development and replaced by the researcher, tour guide and innovator of technology and ecology. In the following, I show how the contestation over territory and commodity flow in the border areas between Peru and Bolivia entail a different but yet related making of frontier space. Here, the intensification of 'border work' (Reeves 2014) both from above and below contributes to form the identities of smugglers and simultaneously their marking as 'other'.

Contested Claims on Space and Identity in the Andes

The highly contested control over commodity flows in the border areas between Peru and Bolivia illustrates how particular kinds of 'frontier spaces' are produced beyond the specific sites of resource extraction themselves. The production of space at the borders also differs from the specific resource extraction sites. For instance, while the sites of resource extraction are often characterized by environmental degradation and the displacement and dispossession of local populations (Li 2015), in the border areas *contrabandistas* (smugglers) and residing populations make claim to the flow of resources and wealth officially controlled by the corporate state. It is another frontier of extractivism, or frontier space, characterized by coexisting and competing regimes of commodity and wealth circulation – and entailing contested claims to sovereignty and control.

The trading practices in these border areas include the smuggling of fuel from Bolivia to Peru, which became widespread when the government of Evo Morales nationalized oil and gas production in Bolivia in 2006 and initiated the subsidization of Bolivian energy prices. Indeed, subsidized energy prices in Bolivia made fuel smuggling into Peru a prosperous venture and – as I have demonstrated elsewhere (e.g. Ødegaard 2016) – also intensified the dissatisfaction with energy policies and price levels among many Peruvians. In contrast to neighbouring Bolivia, Peru has opted for deepening its neoliberal orientation, initiated with Alberto Fujimori's policy reforms in the 1990s, through an emphasis on market liberalization and free trade agreements; it has implemented tax, royalty and policy regimes that accommodate for privatization and foreign investments, especially in the extractive industries. Recent governments in the country have framed these policies within a master narrative of development and progress facilitated by extractivism, presented as key to economic growth and poverty reduction, despite increased conflicts over extractivist projects in recent years. Indeed, the societal and environmental consequences of extractive projects have been increasingly contested through local as well as nationwide protests and demonstrations, intensifying disputes over landscapes and resources, spatial identities and not least of indigeneity in the country (Li 2015). Extractivist conflicts have simultaneously played out by building upon ethnic and regional tensions and stereotypes, reinforcing the dominant and often degrading rhetoric about the Indigenous in Peru, portrayed as backward-looking, second-grade citizens and considered as part of a past temporality, or 'incapable of modernity' (see, for example, Ødegaard and Rivera Andía 2019).

Among the contested energy policies in Peru is an international export agreement concerning natural gas, which has resulted in a reduction of export prices – and at the same time an increase in domestic energy prices (see Ødegaard 2016 for further details). Since then, issues of globalized energy price mechanisms have been a concern, especially among the unprivileged in Peru, such as among Quechua- or Aymara-speaking vendors and *contrabandistas* – with whom my case is concerned. To them, Peru's export of natural gas amounts to 'robbery' of the citizenry, as they see the authorities as illegitimately selling off the country's resources to the disadvantage of the common Peruvian. In this light, the vendors and *contrabandistas* consider the fuel smuggling from Bolivia a legitimate undertaking, claiming that they should also have a right to earn a living by trading in fuel and other commodities. As the *contrabandistas* thus make highly contentious claims to space and the flow of commodities and wealth in the border areas, they can be seen as destabilizing and contesting not only state-corporate control but also the hegemonic envisioning of what constitutes the 'permitted Indian'[13] (Hale and Millamán 2004), which I return to.

The people who bring fuel and other contraband goods from Bolivia to Peru most often self-identify as *contrabandistas* – and proudly so, as they are considered hard-working people providing a valuable social service. They generally travel every week to buy and distribute merchandise, and they are often women, although there are also men involved. The value of hard work is a central dimension of social identities in this context, for women as well as men, partly reflecting the ideological legacy of the colonial experience in the Andes. Among retailers in cities like Arequipa, where I conducted most of my fieldwork, the *contrabandistas* are considered important suppliers of merchandise, and some *contrabandistas* also have their own market pitches there. While some only earn cash every now and then, others run successful businesses based on long careers obtaining contraband goods, often facilitated by the payment of bribes – or the randomness of official interference. Indeed, the delineated border between Peru and Bolivia is difficult to control not only because of the demanding topography but also because of the way in which border trade is socially and spatially embedded in networks of kin, colleagues and cooperating inhabitants in the border communities (see Ødegaard 2016). One of the customs agents I interviewed described the border area between Peru and Bolivia thusly: 'It has been taken by the inhabitants … It is difficult to carry out interferences there, since people become aggressive and defend in any way they can the continued functioning of the place' (my translation). While some communities in these border areas have acquired specific rights as inhabitants of Indigenous territory and others have not, it is worth noting that the trading routes of *contrabandistas* are actualized in areas often considered as *tierras de nadie* (no man's land), referring to places characterized by a history of marginalization, the ambiguous presence of state actors and strong claims to local autonomy. The *contrabandistas'* trading practices thus take place in historically marginalized border areas constituted as the 'last frontier' of the extractivist state.

The town of Desaguadero is a particularly important location for the transport and distribution of contraband goods, being located just at the delineated border between Peru and Bolivia. In Desaguadero one officially crosses the border via a bridge – on foot or by bicycle trolley – or by boat, for those bringing back bigger quantities. In the area around the bridge, one can encounter people who offer the service of bringing your goods across, charging around two soles for the service. Referred to in the media as *pulgas* or *hormigas* (lice or ants), they facilitate contraband trade by bringing customers' goods across the border, sometimes passing many times a day. Desaguadero is often referred to as *tierra de nadie*, and people come here from all across Southern Peru to buy contraband goods, including fuel. In Peru, the fuel is used for ordinary consumption and informal mining.[14] Cross-border trade and fuel smuggling in these areas must be seen as born out of the structural inequalities in Peruvian society and the intersections between class and a racialized hierarchy. These practices can therefore be understood as constituted by deep-seated histories of inequalities but also long-standing practices of exchange and mobility among the Quechua and Aymara (see Murra 1980). Indeed, the *contrabandistas'* trading practices rely on relations of cooperation and exchange between kin, fellow *contrabandistas*, vendors and drivers, as well as inhabitants in the border areas. These relations include ritual exchanges and practices of gift giving, co-parenthood and offerings to the powerful landscape.[15] As *contrabandistas* thus distribute profit by sharing and making investments in kin, colleagues and the powerful surroundings, they can be seen as redirecting the official distribution of fossil fuels and subsuming fossil wealth to other modes of sociality. At this crossroads between fuel smuggling and intensified state interventions, another frontier space is thus produced, where contested claims over control of the commodity flow create the backdrop for a particular kind of persona: the Quechua- and Aymara-speaking *contrabandista*, who does not live up to the conventional stereotype of the Indigenous, but still being constructed as the negative temporal other of the hegemonic citizen.

During the last two decades, Bolivian authorities have intensified measures to limit illicit cross-border trade, by strengthening and modernizing border control systems, raising sentences and militarizing the border, in order to reduce the smuggling of fuel and prevent the national economy from haemorrhaging. The trafficking of fuel not only represents economic loss[16] for countries like Bolivia, where energy prices are subsidized, but also a questioning of state borders, sovereignty and state-corporate interest and investment. The smuggling makes fuel into a petty commodity – and its circulation is informalized in ways that challenge state sovereign claims on its extraction and distribution. Measures have also been introduced on the Peruvian side of the border to reduce smuggling, although not targeted specifically at fuel. Peru's strengthening of controls during the last decade was initiated largely in response to a Trade Promotion Agreement signed with the United States in 2009. Part of this agreement asserts that the United States shall assist Peru in limiting the export of narcotics, modernizing the equipment and procedures of the National Police and Customs Agency and strengthening the rule of law. This intensification of 'border work' on both sides of the border contributes to (re)producing the border areas as a frontier space through contested claims of sovereignty and control over commodity flows, including fuel. Through measures to limit smuggling, the efforts to demarcate the delineated border ('border work') have multiplied, which affects the production of space and identities in areas envisioned as marginal to the nation state. Meanwhile, the *contrabandistas* have responded to intensified state control with further tactics to evade and oppose interventions.

Müller (2021) has explored how the encounters between state actors and *contrabandistas* on the Bolivian side of the border can be understood as 'border spectacles', demonstrating state presence and sovereignty – and often facilitated by media reports emphasizing the drama and force of state intervention. By portraying the protests put up by *contrabandistas* and residing populations in often negative and intimidating ways, Müller notes how media reports further delegitimize cross-border trade as illegal and represent the *contrabandistas* as betrayers of national economic sovereignty. These portrayals correspond with media reports in Peru, which associate *contrabandistas* and vendors with chaos and illegality, crime and violence, and suggest that they are even in possession of arms. Through their involvement in cross-border trade, the *contrabandistas* can be thought of as simultaneously conveying *and* resisting the intersections between class, race and gender in Andean countries, as they are ambiguously positioned between racialized, class-based and gendered categories such as Indigenous and mestizo, rural and urban, poor and rich, private and public: maintaining their rural ways and relations while making a living through trade and entrepreneurship at the margins of the formal economy, and sometimes even running prosperous businesses. Consequently, they are portrayed in media and official discourse as 'another kind of other' – not as 'noble native' or exploited labour reservoir, but rather a symbolically and ideologically ambiguous figure, who has also been increasingly criminalized in recent years (Ødegaard 2016; see also Li 2015). In this regard, the *contrabandista* is made into an example of the somewhat anachronistic personae that global capitalism and extractivism may produce, disturbing the developmentalist narrative and dichotomies of modern versus Indigenous and capitalist versus traditional and destabilizing dominant stereotypes and expectations of the Quechua and Aymara. Thusly, they are continuously being reimagined as 'other', through portrayals and imageries emphasizing their apparent liking for illegality, protest and violence.

The border areas between Peru and Bolivia are constituted as *tierra de nadie* in a way that has enabled and partly legitimized cross-border trade. It is a frontier space where increased state interventions can be seen as spectacles of the extractivist state's presence and force – accompanied by the externalization and 'othering' of *contrabandistas*. This externalization in fact appears as a central dimension of state demonstrations of presence and sovereignty. The Svalbard archipelago is constituted as a frontier space in a different but related way: the Svalbard Treaty affirms Norwegian sovereignty while maintaining the equal rights of countries to operate there. This makes Svalbard not a '*tierra de nadie*' but a territory for the spectacle of marking nation-state presence, and increasingly so. While mining operations and settlements have served this purpose historically, initiatives for a green transition alter the emphasis on mining to environmental management and innovation; from miners to 'innovative subjects'. The 'returning' in Svea is therefore not only a spectacle of coal mining's termination in the archipelago but also marks the presence and sovereignty of the Norwegian state by and through the proclaimed post-extractivist direction laid out for Svalbard. It marks the remaking of Svalbard as site of environmental solutions, a shift that plays on temporal visions of new subjects and externalizes miners as out-of-time survivals. This intensification of questions about presence in frontier spaces thus characterizes both Svalbard and the Andean border areas, demonstrating the production of space and identities in extractivist as well as post-extractivist endeavours. Constituted as spaces of exception – through partly spectacular demonstrations of state presence and control – they are made part of official narratives of (post-)extractivism in ways that may also remove attention from other, conflictual dimensions of (post-)extractivism (e.g. trade agreements, or continued extractivism despite the official environmental narrative). In this regard, the making of anachronistic personae can be considered part and parcel of the spatial and temporal makings of (post-)extractivist frontier spaces.

Conclusions

While extractivist policies create particular frames for the spatial and temporal constitution of frontier spaces, so do post-extractivist policies and narratives. Envisioned as the outskirt frontier of Norway, Svalbard is being made into an environmental showcase through the dismantling of coal mines and the 'return' to nature, including a range of initiatives to facilitate a green transition, making Svalbard also a test case for the future. The case of Svalbard in this regard bears significant differences but also some interesting commonalities with my material from the Andes. Here, the border areas are remade as a contested frontier space not through an official environmentalist emphasis on nature, environment and climate as in Svalbard but through the exceptional form and spectacle of interference by the extractivist state – as well as the externalization, criminalization and 'othering' of the persona of the *contrabandista*, envisioned as betraying the official project of state-corporate growth and development. In this regard, the fuel-trading practices of *contrabandistas* contest the legitimacy of state-corporate extractivism while taking place far away from the extractivist sites themselves. The cross-border trading routes of *contrabandistas* are, in many ways, spaces of exception for the extractivist hegemony of the state, as the circulation and distribution of fuel is informalized – though the state seeks to bring it under their control. Here, another 'frontier space' is thus (re)produced, one in which contested demonstrations of presence and control take place. The *contrabandista* in many ways represents the anachronistic persona of extractivism, as 'out of place and out of time' – they do not correspond to the stereotype of the Indigenous or the enslaved mineworkers as imagined by European engravers in the sixteenth century (see Introduction). In Svalbard, the portrayals of miners illustrate the production of a different anachronistic persona. In the remaking of Svalbard from resource frontier to 'nature's frontier', the miner can only be accounted for by being marked as 'out of time' – and hence made central to narratives of post-extractivism. To put it somewhat crudely, the miner is made into the non-native 'Indigenous' of Svalbard, so to speak, marked as out of time and yet essential for representations of the place. Both of these cases demonstrate the anachronistic personae and spaces that (post-)extractivism may produce, albeit in different ways. In this chapter, I have therefore argued that the analysis of these anachronistic spaces and identities can provide important understandings of the contentions and contradictions, connections and disconnections, of the political economy of (post-)extractivism.

Acknowledgements

I wish to thank research participants and friends in Arequipa, Peru and Svalbard for so generously sharing their perspectives and experiences with me. Some of the material in this chapter also features in previously published works: 'Returning to Nature: Postcarbon Utopias in Svalbard, Norway' (*Social Analysis*, 2022) and 'Border Multiplicities: At the Crossroads Between Regulation and Improvisation in the Andes' (*Journal of Borderlands Studies*, 2016).

Cecilie Vindal Ødegaard is Professor of social anthropology at the University of Bergen. Her research areas include the Andes (Peru), the Arctic (Svalbard) and Norway, with a thematic interest in (post) carbon politics, labour, technology, indigeneity, nature practices and territorial frontiers. Ødegaard is co-editor of *Indigenous Life Projects and Extractivism: Ethnographies from South America* (Palgrave 2019, with Rivera Andía) and *Espacios Transfronterizos de los Andes* (Cross-border Spaces of the Andes, special issue of *Diálogo Andino* 2021, with Müllller). She is author of *Mobility, Markets and Indigenous Socialities* (Ashgate 2010), as well as research articles in *Social Analysis, Journal of the Royal Anthropological Institute, Ethnos, Journal of History and Anthropology, Journal of Development Studies, Journal of Borderlands Studies, Journal of Ethnobiology and Ethnomedicine,* and more.

Notes

1. 'Post-extractivism' here is not meant to imply the end of extractivism but refers to initiatives for post-carbon transition.
2. Fieldwork periods in Peru were between 1997 and 2016; and in Svalbard since 2019.
3. To emphasize narratives in this way is not a claim about their truthfulness but a way to explore their effects.
4. To emphasize Svalbard as an environmental showcase does not mean that Norway cannot also be categorized as an 'extractivist state' – it can, and increasingly so.
5. For purposes of argument, the emphasis here is more on the narrative and discursive makings of social identities and categories, rather than the more complex makings and negotiations of individual subjectivities.
6. Briefly because I have already published widely on my work in Peru.
7. Most mineworkers in Svalbard have been men, although some women were also included from the 1980s onwards.

8. Except for the levelling to the ground of mining structures in Pyramiden and other mining communities in Svalbard by British-Canadian forces in 1938–39, to prevent their exploitation by Hitler (Fløgstad 2007: 90).
9. Although the discussions about closing Gruve 7 were reopened with increased energy prices and Russia's war on Ukraine.
10. The Longyearbyen Local Board finalized a new energy plan in 2023, as requested by the government.
11. Coal mining has been Store Norske's core activity since its establishment, in addition to the management of state land and buildings. The company is owned by the Norwegian state.
12. A term coined by Dina Brode-Roger; personal communication.
13. Coined by Silvia Rivera Cusicanqui, the notion 'permitted Indian' pinpoints how neoliberal capitalism, as a cultural project, contributes to the rising prominence of Indigenous voices while simultaneously limiting their transformative aspirations (Hale and Millamán 2004: 17).
14. Along Cuenca Suche, the river that goes through Desaguadero, there are a range of un-authorized mining businesses.
15. Including non-human beings of the landscape, such as *pachamama* (earth) and *apus* (mountains).
16. In 2011, it was stated that illegal export of fuel amounted to a loss of approximately $450 million US dollars per year (Ødegaard 2016).

References

Büscher, Bran, and Robert Fletcher. 2020. *The Conservation Revolution: Radical Ideas for Saving Nature Beyond the Anthropocene*. London: Verso Books

Cowie, Jefferson, and Joseph Heathcott (eds). 2003. *Beyond the Ruins: The Meanings of Deindustrialization*. Ithaca: Cornell University Press.

Dodds, Klaus, and Mark Nuttall. 2016. *The Scramble for the Poles*. Cambridge: Polity Press.

Ferry, Elizabeth E., and Mandana E. Limbert (eds). 2008. *Timely Assets: The Politics of Resources and Their Temporalities*. Santa Fe: School for Advanced Research Press.

Fløgstad Kjartan. 2007. *Pyramiden: Portrett av ein forlaten utopi*. Spartacus Forlag.

Franquesa, Jaume. 2018. *Power Struggles: Dignity, Value, and the Renewable Energy Frontier in Spain*. Bloomington: Indiana University Press.

Gerlach, Julia, and Nadir Kinossian. 2016. 'Cultural Landscape of the Arctic: Recycling of Soviet Imagery in the Russian Settlement of Barentsburg (Svalbard, Norway)', *Polar Geography* 39(1): 1–9.

Hale, Charles, and R. Millamán. 2004. 'Rethinking Indigenous Politics in the Era of the "Indio Permitido"', *NACLA Report on the Americas* 38(2): 16–21.

High, Mette, and Jessica Smith. 2019. 'Introduction: The Ethical Constitution of Energy Dilemmas', *Journal of the Royal Anthropological Institute* 25(S1): 9–28.

Hilgers, Mathieu. 2010. 'The Three Anthropological Approaches to Neoliberalism', *International Social Science Journal* 21(202): 351–64.

Igoe, James. 2005. 'Global Indigenism and Spaceship Earth: Convergence, Space, and Re-entry Friction', *Globalizations* 2: 1–13.

Li, Fabiana. 2015. *Unearthing Conflict: Corporate Mining, Activism, and expertise in Peru*. Durham: Duke University Press.

Meyer, Judith. 1996. *The Spirit of Yellowstone: The Cultural Evolution of a National Park*. London: Rowman and Littlefield.

Muehlebach, Andrea. 2011. 'On Affective Labor in Post-Fordist Italy', *Cultural Anthropology* 26(1): 59–82.

Müller, Juliane. 2021. 'Encuentros y espectáculos fronterizos: Comercio transfronterizo y disputas por la legitimidad y riqueza en el occidente de Oruro, Bolivia', *Diálogo Andino* 66: 299–311.

Murra, John Victor. 1980. *The Economic Organization of the Inca State*. Greenwich: Jai Press.

Ødegaard, Cecilie Vindal. 2016. 'Border Multiplicities: At the Crossroads Between Regulation and Improvisation in the Andes', *Journal of Borderlands Studies* 31(1): 23–38.

———. 2022. 'Returning to Nature: Postcarbon Utopias in Svalbard, Norway', *Social Analysis* 66(2): 1–22.

Ødegaard, Cecilie Vindal, and Juan Javier Rivera Andía (eds). 2019. *Indigenous Life Projects and Extractivism: Ethnographies from South America*. London: Palgrave Macmillan.

Rasmussen, Mattias Borg, and Christian Lund. 2018. 'Reconfiguring Frontier Spaces: The Territorialization of Resource Control', *World Development* 101: 388–99.

Reeves, Madeleine. 2014. *Border Work: Spatial Lives of the State in Rural Central Asia*. Ithaca: Cornell University Press.

Rose, Nikolas. 2007. *The Politics of Life Itself: Biomedicine, Power, and Subjectivities in the Twenty-First Century*. Princeton: Princeton University Press.

Saville, Samantha. 2019. 'Locating Value(s) in Political Ecologies of Knowledge: The East Svalbard Management Plan', in S.M. Saville and G. Hoskins (eds), *Locating Value: Theory, Application and Critique*. London: Routledge, pp. 173–85.

Sokolickova, Zdenka, and Thomas Hylland Eriksen. 2022. 'Extraction Cultures in Svalbard: From Mining Coal to Mining Knowledge and Memories', in Sverker Sorlin (ed.), *Resource Extraction and Arctic Communities: The New Extractivist Paradigm*. Cambridge: Cambridge University Press, pp. 66–86.

Totland, Per Arne. 2016. *Kaldfront: Konfliktområdet Svalbard gjennom 100 år*. Oslo: Cappelen Damm.

Tsing, Anna L. 2003. 'Natural Resources and Capitalist Frontiers', *Economic and Political Weekly* 38(48): 5100–6.

———. 2005. *Friction: An Ethnography of Global Connection*. Princeton and Oxford: Princeton University Press.

———. 2015. *The Mushroom at the End of the World: On the Possibility of Life in Capitalist Ruins*. Princeton: Princeton University Press.

West, Paige, James Igoe and Dan Brockington. 2006. 'Parks and Peoples: The Social Impacts of Protected Areas', *Annual Review of Anthropology* 35: 251–77.

Whitesell, Edward A. 1996. 'Local Struggles Over Rain-Forest Conservation in Alaska and Amazonia', *Geography Review* 86(3): 414–36.

Winkler-Reid, Sarah. 2017. '"Doing Your Best" in a London Secondary School: Valuing, Caring and Thinking Through Neoliberalism', *The Sociological Review* 65(1): 137–53.

— Chapter 2 —

TECHNOLOGICAL PROMISES OF 'GREEN' EXTRACTIVISM IN HISTORICAL MINESCAPES

Narratives and Materiality of the Mining Revival in Andalusian Wastelands

Doris Buu-Sao

Discourses of 'sustainable mining' are surprisingly spreading in some historical mining territories of Europe and United States. Coming from the Global South, where corporations appropriate sustainable development language as a legitimizing strategy (Kirsch 2010), they are now reaching Northern rural, often deindustrialized areas. There, a mining revival appears as a paradoxical remedy to the economic but also ecological crisis: the 'energy transition' is presented as depending on mineral resources necessary to build solar panels, wind turbines or electric cars. In 2009, the European Commission presented the Raw Material Initiative to the Parliament: it called for the reactivation of extractive industry, as a 'driver of economic growth in some more remote European regions',[1] which would contribute to secure the access of European industry to raw materials. Since then, new mining projects have proliferated across rural peripheries of the continent (Buu-Sao 2021b; Del Mármol and Vaccaro 2020).

Spain is a pioneer territory in this regard, especially its Southern region of Andalusia. Five years after the 2008 crisis, more than one third of the Andalusian working population was unemployed (36.22 per cent), ten points above the national rate (26.10 per cent). Facing those long-lasting effects of the financial crisis, in 2014, 'Andalusian authorities made a political decision to take advantage of all the resources of the region'.[2] Andalusian deposits had been mined for thousands of years, especially in the Iberian Pyrite Belt, a geological area very rich in non-ferrous metals (mainly copper, zinc and gold). But the mining industry had been dismantled, along with other sectors, following the 'industrial modernisation' reforms of the 1980s (Marín Arce 2006). Political support of the rebirth of this sector gave rise to a mining recovery programme, which is outlined in the *Andalusia Mining Strategy 2020*, a policy document released in 2015. Ever since, discourses, policies and extractive projects in favour of a mining revival have spread in the region, which hosts seven operative metallic mines and a dozen exploration or projected mining permits.

Extractivism, a concept initially coined in Latin America, describes a regime of natural resource extraction that serves as the driving force of economic capitalism (Gudynas 2019). Copper mining clearly falls into this category. Although the mining industry is an emblematic case of 'conventional' extractivism, I propose to tackle its revival in Andalusia in terms of a 'green' extractivism. The revival of the mining industry is based on justification schemes that highlight the need for rare earths, 'critical' metals (such as lithium or tungsten), as well as 'traditional' ones (such as copper) in the fight against climate change. Beyond the justifying logic of this 'green energy bargain' that associates decarbonization with mining (Phadke 2018), energy transition policies feed economic projections that favour a rise in metal prices and, consequently, greater profitability of mining projects. The 'renewable energy-extraction nexus' is at the heart of analyses of 'green' extractivism, both in its discursive and economic-financial dimensions (Dunlap and Jakobsen 2020). How is a mining revival made feasible and profitable, in a European context where the history of deindustrialization, the contentious potential of environmental criticism and the low grade of minerals mined for millennia put these projects at risk? This chapter explores the extent to which technological promises solve these paradoxes of 'green' extractivism. The next section develops its theoretical and empirical basis, before opening on the ethnography of the mining revival in Andalusia.[3]

Studying 'Green' Extractivism in European Wasteland

My ethnography of the 'green' extractive economy in Andalusia draws on the sociology of science and technology, of public action and on the anthropology of mining. According to the sociology of 'technological promises' (Van Lente 2017), images of the future shape decisions made in the present. Jens Beckert underlines the performative value of economic projections that he calls 'fictional anticipations'. They 'shape anticipations and provide justifications for decisions, reducing any possible disorientation of decision-makers given the openness of the future' (Beckert 2013: 235). The narrative of a future characterized by a growing demand for raw materials, on the one hand, and the vitality of a sector able to create employment in regions durably affected by multiple crises, on the other, acts as an adjunct to the mining revival. The 'closure of futures' is thus based on predictions and promises that are audible in the dominant language, excluding alternative futures (Brown, Rappert and Webster 2017). These 'future scenarios' (ibid.) rely on the production of 'socio-technical imaginaries', defined as 'collectively held, institutionally stabilized, and publicly performed visions of desirable futures, animated by shared understandings of forms of social life and social order attainable through, and supportive of, advances in science and technology' (Jasanoff and Kim 2015: 4). But the analysis of promises also leads to questions of the materialization of these imaginaries into economic activities vulnerable to criticism. Technological promises 'are not only a matter of discourse and representations; they also involve practices of exploration and experimentation; they are linked to the investment, mobilisation, circulation and accumulation of resources' (Joly 2010: 204). These materializations occur thanks to the production of the credibility and legitimacy of technological promises.

In order to unveil the symbolic and material production of technological promises, one needs to go beyond the historically and daily instituted symbolic frontier between the Economy and the State as distinct spheres (Mitchell 1999): although referring to industrial projects, technological promises deeply rely on public policy. A Green Keynesian paradigm heralds the active role of 'entrepreneurial states' in fostering the 'continued and accelerated development of commercially focused, privately developed green technologies' (Goldstein and Tyfield 2018: 76). I see this 'entrepreneurial state' as not only injecting capital into innovation for green technologies but also striving, in parallel, to set the legal and discursive framework that enables and legitimizes industrial activities aimed at 'fixing' rather than 'shifting' from the dominant socio-economic paradigm (Bina 2013) – which, in other words, gives rise to a 'non-disruptive disruption' (Goldstein 2018). In analysing this public and private co-production of socio-technic and sociopolitical change (Jasanoff and Kim 2015), this chapter pays special attention to the mythical dimension of the promises of a mining revival. 'Policy myths' contribute to the re-enchantment of public action by promoting miracle solutions that are often vague and imprecise (Desage and Godard 2005). The constitutive ambiguity of policy myths favours the gathering of various initiatives and actors, but it does not erase competing interests.

One of the recurring myths surrounding the promise of the mining industry is the emphasis on territories that are considered unproductive, whose resources are waiting to be tapped. In this process, territories are qualified as 'waste', a category that covers both a negative dimension, what is superfluous, and a positive dimension, what can be valued in the capitalist sense of the term (Franquesa 2018). Jesse Goldstein has shown how eco-modern promises for clean tech and green growth renew the modern project of 'improving' unproductive, 'waste' land that transforms nature into a *terra economica*, conceived as a passive repository of resources to be extracted (Goldstein 2018). Likewise, 'green' extractivism reproduces the production of 'sacrificial zones', which are 'othered' territories from which fossil capitalism extracts value while confining industrial waste and contamination (Andreucci and Zografos 2022).

Andalusia and its historical Riotinto copper mine are a privileged observation site of the technological promises of 'green' extractivism in territories built as wastelands. This Southern region, at the crossroads of Mediterranean and then colonial trade after the military conquest of America, was initially at the forefront of the development of an extractivist capitalism controlled by the bourgeois and noble local elite. These politico-institutional arrangements have had long-lasting effects on Andalusia, which has become a dependent and impoverished periphery of Spanish capitalism (Arenas Posadas 2016). Since these arrangements took place in the context of the *Reconquista* of a racialized region, Al-Andalus, some historians consider Andalusia to be shaped by a colonial history that today translates into the racialization of its population through the archetype of the lazy Andalusian – 'Andaluz vago' (García Fernández 2019). The 2008 crisis violently hit Andalusia, and the project of 'green' extractivism, from new mines to multiple solar power plants, can be understood as the reactualization of this historical 'othering' (Andreucci and Zografos 2022) of the region. At the turn of the nineteenth century, this millenary mine, exploited by a British company following a 'colonial enclave' model (Ruiz Ballesteros 1998), was the first producer of copper in the world (Chastagnaret 2000). Since the 1980s, its inhabitants have experienced the harsh impact of deindustrialization (Arenas Posadas 2017) and still suffer from structural unemployment – in 2015, between 23 and 41 per cent of the working

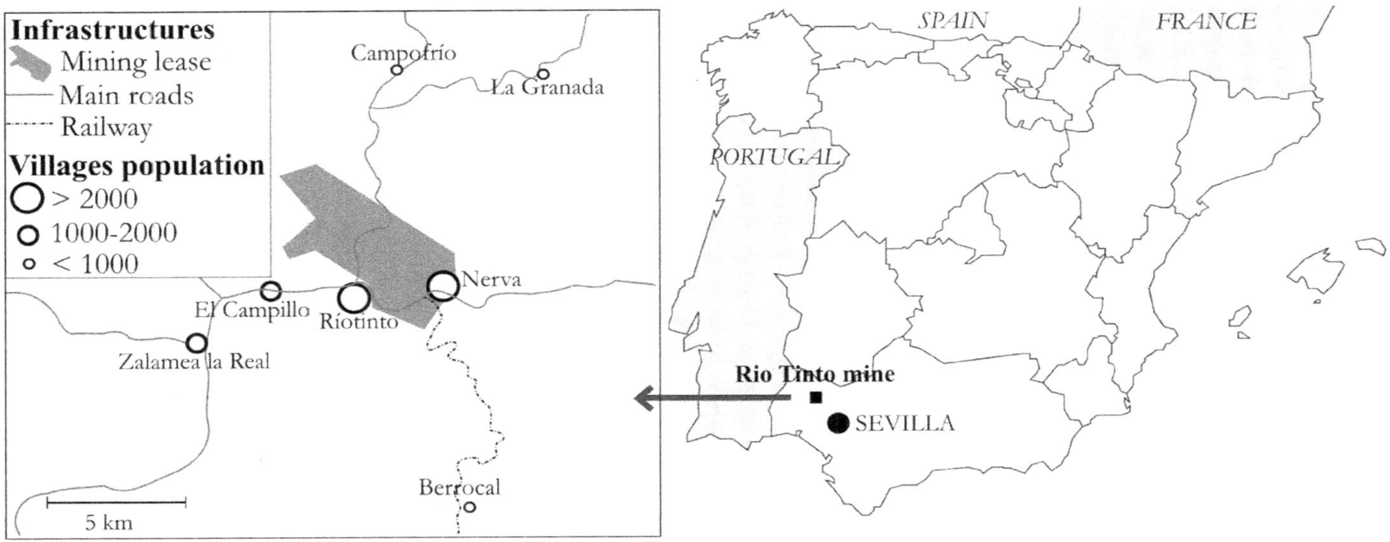

Figure 2.1. Localization of mining revival encounters. © Doris Buu-Sao

population of the seven villages that comprise the area were unemployed. In 2015, the Riotinto mine resumed production after fourteen years of inactivity. While its museum is a famous example of mining heritage making (Hernández Ramírez and Ruiz Ballesteros 2007), corporations and authorities now aim to make Riotinto a model of twenty-first century mining.[4] This case articulates processes regarding mining heritage, industrial waste management and promises of mining revival, which make it particularly relevant for investigating how multiple 'mining temporalities' (D'Angelo and Pijpers 2018) mingle in the materialization of 'green' extractivism.

Inside a 'Professional Village' of Future Mining

In October 2019, Sevilla hosted the third Mining and Minerals Hall*[5] (MMH) at the Palacio de Exposiciones y Congresos. The Association of Andalusian mining enterprises (AMINER) co-organized the event with the Junta de Andalucía (Regional Government of Andalusia), which acted as a 'promoting' entity. For three days, a micro mining world materialized. To the extent that trade fairs are large-scale professional gatherings 'where entire industries emerge', in the sense that they come 'into existence *and* become visible' (Leivestad and Nyqvist 2017: 2), the 'professional village' (ibid.) that was temporarily created was a rich empirical site to track the production of discourses on the future of mining and their tangible effects.

Theatres of Optimism

Reaching the venue, convenors crossed a vast circular area where they passed between several extraction machines before entering the central lobby. After they gave their name at the reception, signed the attendance sheet and paid the 350 € fee if they had not done it online, participants were handed a badge and a customized MMH backpack. According to the 'scientific programme' that was inside the bag, participants could choose between two different auditoriums where talks would be simultaneously delivered. In the main auditorium, which had been named after a mining company who was operating in the region and who was the main event sponsor, eight seminars and as many roundtables had been schedule on topics such as 'social licence to operate', 'biodiversity', 'circular economy', 'energy efficiency' and 'innovation in mining industry'. In the smaller auditorium, six 'technical seminars' took place in parallel; these were conducted by members of the Horizon 2020 research projects related to mining extraction. They presented technologies that would help to extract more, in a cheaper, safer and more environmentally friendly way. From the first morning to the third afternoon of the event, seventy convenors delivered speeches and answered the moderator or public's questions.

Throughout all these talks, roundtables and technical seminaries, optimism was repeatedly expressed. Mining industry officials stated that the urbanization of developing countries and the global need for energy transition were already fuelling a growing demand in metals; mining companies could contribute to satisfying this need and extract European natural resources. Experts from the European Commission highlighted the dynamism of research centres and public-private networks for mining innovation; they underlined the will of EU to support them, 'not only for investigation, but also in order to bring innovation into the reality'.[6] Scientists and engineers of the Horizon 2020 research projects outlined the positive outcomes of their work and future perspectives regarding the standardization of technological innovations for 'sustainable mining'. Civil servants of the regional government claimed that a mining revival and the development of the entire value chain, from extraction to metallurgy, would boost Spanish growth, create jobs and help remedy the depopulation of rural regions. Members of the Spanish administration announced programmes of 'modernization, simplification, and digitalization', a recurring conceptual trio associated with the reform of mining licences and administrative processing. The administration was blamed for being one of the main obstacles to mining recovery, reflected in the many jokes about the inefficiency of Spanish bureaucracy during lunches and coffee breaks.

The president of Euromines (the association representing mining companies' interests in Brussels) was emblematic of such optimistic stances. He explained during a roundtable called 'Current situation, threats and opportunities for metallic mining': 'Climate change is pushing change towards electrification. To do that, Europe is going to explore more, to produce more. We won't be able to produce this change without traditional materials*.' Not only was global warming turned into a business opportunity, but the Global South's growing capacity to resist the monopolization of extraction by Western companies was opening new markets: 'Decolonization is our opportunity. I'm extremely optimistic, we are going to do this', added the president of Euromines. Nobody contradicted such repeated assertions of optimism, contributing to the production of a coherent narrative on the mining future.

Although the fourth MMH was postponed to 2022 due to the Covid-19 pandemic, optimistic speeches still saturated the public narrative on the mining revival in Andalusia. In Sevilla, in October 2021, the Spanish Superior Council of Scientific Research (CSIC) inaugurated the 'European Institute of Innovation and Technology on Raw Materials (EIT-RM) Hub Southern Spain'. The EIT-RM is a consortium whose mission is 'to enable sustainable competitiveness of the European minerals, metals and materials sector along the value chain by driving innovation, education, and entrepreneurship'.[7] Its seven regional hubs support the creation of networks and programmes designed to promote sustainable mining. The inauguration of the Spanish Hub in the Casa de la Ciencia (House of Science) Museum was open to members of the public who had signed up, which was checked at the entrance. Some twenty participants – mainly delegates of public administrations and mining companies – sat in the auditorium to listen to the presentations offered by members of the CSIC, of European of research and innovation programmes and of the Junta de Andalucía. Speaker after speaker echoed the necessity of diversifying supply sources for the energy transition and the importance of unlocking 'the mineral potential of Southern Spain' through 'bringing together academia, industry, and research institutions'.[8] Though the pandemic revealed the dependency of European industry on foreign supply, this has not prevented Andalusian mines from maintaining production, they claimed. Even more, in the 'post-pandemic scenario' and the need to boost the economy, the development of mining extraction was again presented as a solution to regional problems.

This pervasive optimism echoes the 'future positive mode' of development programmes (Mosse 2005), in which positive views of the future effects of such projects are key to their implementation. As David Mosse argues, the materialization of development projects relies on a subtle balance. Development projects depend on a diversity of actors through a process of coalition-building that contains diverse and sometimes irreconcilable perspectives while needing to persuade a broader audience through the production of a coherent discursive frame. Hope and optimism are key to achieving such balance – that is, drawing an imagined future in which development projects will fulfil their promises. Similarly, European and Andalusian civil servants claim that mining recovery in Southern Europe will solve multiple problems, from climate change to economic crises, aggregating a diversity of actors and interests while creating a master frame that is appealing to the public, in the process relying on a 'strategic usage of temporal frames' (Luning 2012a: 208) characterized by a 'historical amnesia' of darker moments of the mining industry, as Sabine Luning witnessed at a Canadian mining convention. In the 'professional village' of the MMH, environmental and social distresses caused by mining were erased from dominant discourses on the glorious Andalusian mining past. Among the multiple accidents, such as the rupture of the Aznalcóllar mine tailings pond in 1998, and social conflicts related to mining history, it is worth mentioning a particularly traumatic episode often presented as the first environmental conflict in history (Martínez-Alier 2002: 59). On 4 February of 1888, thousands of farmers, inhabitants and workers of the Riotinto copper mine protested a highly polluting technique of purifying metal called 'open-air calcination'; they were violently repressed by the army, who killed several hundred people (Chastagnaret 2017). One can note that the history of the policy of industrial dismantling implemented by the Spanish government in the 1980s has been silenced. Such a policy led to the closure of historical metallic mines and the unemployment of thousands of miners in the region (Marín Arce 2006).

Although many negotiations and decisions are taken within more discrete arenas, public events where discourses are coherently framed should not be reduced to mere institutional façades. Backstage, public and private actors of the mining sector meet, exchange ideas and eventually contribute to the circulation of public mining revival narratives. Such events can be thought of as 'theatres of optimism', echoing the international conferences on corporate social responsibility, which Dinah Rajak conceptualizes as 'theatres of virtue'. In her view, those ritualized spaces host performative discourses that establish CSR as a development orthodoxy, through the construction of coherent narratives and the silencing of critical voices (Rajak 2011). Similarly, public events such as MMH, or the establishment of the EIT-RM Hub Southern Spain, constitute theatres of optimism that display 'the appearance of consensus between participants, and the allegiance to the shared goals' (ibid.: 13).

Materializing 'Sustainable Mining' and Its Partnerships

The optimistic frame has productive effects, thanks to the logic of enrolment and persuasion (Mosse 2005). The extractive industry relies on a positive vision of the future, which can shape decisions made in the present: natural resource commodification depends on predictions of volume, economic value, extraction cost, social acceptation, etc. (Laurent and Merlin 2021; Luning 2018). Therefore, optimistic discourses are not just part of a collective performance aiming to build a coherent and bright face for the outside: the construction of a desirable future has tangible effects on present reality.

During the 2019 MMH public presentations, there were never more than a hundred persons, while organizers claimed that 700 to 800 persons attended: many convenors spent most of their time at the commercial fair that took place in another section of the Palacio de exposiciones y congresos, in two large halls, where more than a hundred organizations held a stand. Although the mining industry is anything but new, novelty was at the core of exhibitors' messages. Spanish, regional and Portuguese administrations promoted new opportunities of investment – for instance, Portugal's vast maritime territory. Mining companies celebrated new projects with extended capacities of operation in time and space, thanks to modern technologies and processes. Manufacturers drew a bright future for more secure, automated and efficient mining thanks to artificial intelligence – as did a company that had installed a control tower labelled 'digital mining in action'.

Exhibitors could be classified into five different categories. First, companies that hold a mining licence, generally in Andalusia, had relatively sober stands – except one that reproduced a TV show set and was called 'The big game of copper'. Second, subcontractors and consultants offered a vast array of services ranging from exploration activities to tasks and machines related to mineral extraction, processing or restoration. They represented most of the exhibitors; their stands used to be the most appealing, with different kinds of machinery, miniatures and, above all, catering services; I saw dozens of people with a glass of wine sat gathered around a *tapa* of Iberic ham. The third type of exhibitors included professional organizations that represent mining enterprises through associations (such as AMINER), networks (such as the Iberian Sustainable Mining Cluster) or specialized reviews. Then came stands of public administration entities and, finally, institutions related to research, innovation and training for the sector. Together, these stands made visible the different segments of the mining economic field (Leivestad and Nyqvist 2017). They materialized the diversity of actors (except for the miners) that make mining extraction possible, unveiling the importance of subcontractors, who now execute most extractive tasks.

One market issue was made especially clear during MMH: mitigation of the environmental impact of the mining industry. During a lunch, I shared a table with engineers working for three different consultant enterprises. The first specialized in energetic efficiency, mainly through solar plant installation; the second in the 3D modelling of environmental impacts of industrial projects; and the third in the restoration of deteriorated environments through reforestation. Each new person arriving to the table asked the typical question – 'which enterprise are you from?' – and engaged in the conversation, commenting on their respective professional activities and looking for common acquaintances. It only took a moment for business cards to be exchanged. As Miguel Angel, the energy efficiency consultant, explained: 'the congress in itself is not so interesting, but I have to come and say hello: networking* is the key'. If socializing is at the core of any trade fair (Leivestad and Nyqvist 2017), in this case, the exchange of information and business cards would have directly contributed to the emergence of 'green mining', a new market segment that was announced in the scientific programme. Consultants were eager to move into the niche created by environmental concerns – 'if [mining companies] want to survive, they have to turn green', said Miguel Angel. Beyond the symbolic stage provided by MMH, informal encounters strengthen professional networks that in turn 'provide social infrastructure for emerging markets and industries' (Aspers and Darr 2011: 774).

Such events are also where more institutionalized partnerships are celebrated. The 2021 inauguration of the Southern Spain Hub ended with repeat invitations to mining companies to join public networks of science and innovation for sustainable raw materials extraction. Symmetrically, the bi-annual organization of MMH is also a call for public administration to take part in the mining revival. The Junta de Andalucía is eager to do so, through its active participation in the organization of the event:

> I really think that the most important contribution is the presence of the administration itself in supporting the sector. … . We have participated in everything, involving other administrations, people from universities, some companies … But I see that as more specific. What really matters is that the administration shows its support to the mining sector, something that has not always been the case throughout history. There have been times when the administration did not understand that it was an important sector in the region. Now, it is different.[9]

In this theatre of optimism, staging governmental support reflects a historical perception of the Spanish central state and regions as not conscious enough of the importance of mining. In his analysis of UN summits on sustainable development, Carl Death highlights the centrality of 'partnership'. The concept works as a powerful discourse that advocates for a broad public and private 'range of self-governing actors as necessary and important for the implementation of sustainable development' (Death 2010: 80). These collaborations produce a 'shift from more centralised and hierarchical forms of governance to voluntary, flexible partnerships … [which] represents a new form of rule at a distance, according to a rationality based on the operation of the market' (ibid.: 86). However, promises of 'green mining' implemented by responsible private investors also depend on the commitment of public administration, as the MMH website states:

> Today's mining industry is moving towards green mining, demonstrating an unavoidable commitment to sustainable and responsible production, … as well as joining the fight against climate change. But this commitment requires a significant investment effort, hence the importance of having the commitment of the public authorities.[10]

The call to 'expedite' mining licensing procedures, repeatedly expressed during public events, is not a mere tagline restricted to corporate actors. Collaboration in the organization of public events leads officers of the mining department to regularly meet with executives of Spanish mining companies, with whom they also share sociological proximity – many of them are mining engineers who studied at the Polytechnic School of Madrid. During these encounters, public and private agents of the mining sector come to share a common narrative regarding the public administration's responsibility. Together they put pressure on the regional government of Andalusia, from inside and outside, to commit to the mining revival. The Regional mining department is currently working on the Sustainable Mining Strategy of Andalusia 2030, a document that will set up the core guidelines for mining policy in the next ten years. The department's chronic staff shortage has led the couple of engineers responsible for the elaboration of the document to appeal to external actors: the mining administration, like other policy sectors, is facing a structural lack of resources that means the delegation of part of its work to private actors. The mining administration and its staff have relied on the guidelines established by the Organisation for Economic Co-operation and Development (OECD) in the report, which was prepared with the support of the Andalusian Department of Mines and AMINER.[11] Private consultants will prepare a first draft that mining companies will be invited to discuss, along with other 'stakeholders*'.[12] Along official partnerships and discrete collaborations, public and private actors of mining sector are working together to design mining revival policy as a technological, de-politicized and market-oriented solution. As Carl Death states, 'partnerships are not neutral, value-free tools, but rather they prioritize certain types of actors, specifically those with technical expertise or project management skills' (Death 2010: 86). In public forums as in private political and economic circles, alternative voices are set apart from those partnerships that establish '"the global market" as the fundamental mechanism through which [sustainable development] can be achieved' (Rajak 2011: 11).

Professional events in which public and private actors engage in enduring collaborations contribute to the symbolic and material production of technological promises of 'green' extractivism. However, the reopening of old mines has not waited for MMH or the new regional mining policy. Since the second half of the 2000s in Andalusia, while private and public actors of the mining sector have been fostering policy reforms, transnational companies have undertaken new exploration and extraction projects. It is now time to turn to an iconic 'twenty-first century mine', which is part of a mining history that has challenged technological promises.

Reopening the Pit

There is a growing body of anthropological literature addressing political, material and symbolic processes related to infrastructure. Conceived as 'built networks that facilitate the flow of goods, people, or ideas and allow for their exchange over space' (Larkin 2013: 328), infrastructure can more broadly refer to any assemblage that stores productive powers of material and human labour for capital accumulation (Boyer 2018). Central to economic production, infrastructures 'have long promised modernity, development, progress, and freedom to people all over the world' (Anand, Gupta and Appel 2018: 3). They connect political programmes, capitalist global processes and localized places. Infrastructures and the promises they convey can then be studied as 'a critical site through which politics is translated from a rationality to a practice, in all its social, material, and political complexity' (ibid.: 20). To this extent, mining infrastructures constitute a rich empirical site to track the uneven materialization of technological promises.

Performances of Innovation

One morning of the MMH week that unfolded in October 2019, I arrived once more to Sevilla's Palacio de exposiciones y congresos. A team of workers was dismantling the machines installed in the circular area: MMH had ended the evening before, but some participants had subscribed to an optional visit to one of three mines in Andalusia. Four men (all were engineers, from different European and American nationalities) and I were welcomed by MMH organizers, who invited us to get in a Volkswagen minibus parked at the entrance. After an hour and a half trip up to the Sierra de Huelva, we arrived at the entrance of the Riotinto mining area. A security guard got in the vehicle and wrote our names on a post-it without asking us to spell them. I was surprised he let us through the gate without any additional identity checks.

A woman wearing a safety vest and a helmet then invited us to follow her into the central office. We entered a meeting room where four men were waiting for us. After one of them delivered a traditional security contact, a man in his fifties presented himself – in English, so everyone could understand him. He was Enrique, the manager of the Riotinto project. The mine was the oldest in Spain, maybe in Europe, he explained: its gold and copper have been exploited since the Roman civilization and even before, in the era of the Tartessos (1200–500 BC). Since its reopening, the main challenge has been to operate the mine in a profitable and secure way. Enrique then put three engineers in charge of the visitors: Leandro, the project manager, Mario, the engineer in charge of the extraction, and Maria, the engineer responsible for the ore process. They drove us to different parts of the mining area while explaining the mining process: exploration of abandoned open pits and underground operations; extraction in the (by then) 150 meters deep Cerro Colorado open pit; ore processing in crushing and grinding machines; ore washing to get a powdered concentrate of copper, which is sold, mainly, to Chinese buyers; storage of mine tailings in a 580-ha pond. They also took us to an archaeological site located inside the concession, where a team was excavating the Roman remains of mining activity – they had found 'treasure' in the form of 115 silver and gold imperial coins. Back to the meeting room, three hours later, the director Enrique was waiting for us with sandwiches, fruits and sodas.

On the road back to Sevilla, the four men said that they had found the visit surprisingly 'transparent' in their experience. When I came back a few days later to the Riotinto village, the engineer Mario offered me a second tour to help me understand better the extraction process. He explained that the mine always welcomes visitors; it is part of his work to drive them to the installations and explain how they operate the mine. Back at the central office, I spoke to Luis, the head of the communication department. He mentioned, within the CSR (Corporate Social Responsibility) strategy, a 'programme of visits'. University students, scholars and also local children were welcomed for one-day visits that focused on contemporary as well as on past mining: 'beyond modernity, the mine is related to [the children's] history and we consider that as something important'. Visits were one of the many CSR guidelines that, Luis told me, aimed to 'bring the mine and the population closer together' while producing positive impacts on the 'community', through actions such as financial support to entrepreneurial projects beyond the mining sector.

Beyond narratives of transparent and responsible industry, these visits showcase the ambivalent relationship between mining temporality and 'green' extractivism. After fifteen years of inactivity, the Riotinto mine, operated by Atalaya, was presented as a brand new one, in line with the discursive frame elaborated in theatres of optimism such as MMH. Maria, the process engineer, was particularly proud to show us the new grinding mill (the 'biggest in Europe', according to her) that was about to be operational; it would help to increase the quantity of processed ore up to 15 million tons a year, adding thirteen years to the lifespan of the mine. The director and the head of the communication department distinguished mining remote history, which they considered a cultural heritage to be preserved and valorised, from the present day and efficient, transparent and responsible mining, which had a long life ahead. Year after year, multiple promises of mining expansion have come from Atalaya Mining company. Its officials endlessly promote a brighter future for copper extraction, which is reliant on a mix of optimistic projections regarding economic trends and geological resources, and technological innovation. In August 2020, Atalaya presented to the Junta de Andalucía a request to substantially modify environmental policy to expand the volume and duration of mining operations. The 'Technical environmental project', one of the documents sent as part of the application, stated that managers planned an increase of the mining area by 41 per cent (from 159 to 225 ha), to mine 479 million tons of additional rock, among which would be 197 tons of mineral and 281 tons of sterile waste. The project was presented as 'a modern mining activity … that meets the highest standards of technology, safety, resource and raw material management and environmental protection, and whose fundamental pillars of sustainability are the environmental and social protection of the area and socioeconomic development over time'.[13] While the application is still being processed, Atalaya Mining has been investing in expanding its activities in other ways. For example, in 2021, it bought twenty exploration permits to the east of the copper mine and in the neighbouring Extremadura region. It also financed a research project on new extraction technology for producing pure ore, led by a woman with a Ph.D. in electrochemistry, who is often described as a 'genius';[14] the company is now investing 12 million euros in a new plant that will use her technique to produce up to 10,000 tons of pure copper or zinc per year. Presenting this 'exciting project' to the press, the Atalaya Mining CEO announced that 'it is a process aligned with the objectives of sustainable development and climate neutrality, which will use renewable energy from the solar plant that we are going to build in Riotinto'.[15] Riotinto mine has been, until now, using only fossil fuels to produce a 22 per cent copper concentrate that is sent by trucks and then by boat to the buyers. The construction of a plant to produce pure copper, powered by solar panels, is a way to extend the value chain, from energy to metal production, in an environmentally friendly way.

At stake, therefore, is mining expansionism in time, space and value chain. Such expansionism relies on technology that can make mining infrastructure compliant with UN Sustainable Development Goals and that appeals to the Andalusian administration, shareholders, local populations and authorities as well. However, expansion projects are part of a techno-economic system involving other extractive logics that impact human and more-than-human nature (Dunlap and Jakobsen 2020). According to the documents presented to the Junta de Andalucía in 2020,[16] the Riotinto expansion project implies the use of 66 per cent more explosives (from 2,100 tons each year to 3,500) and fossil fuels (from 15,993 million litres to 26,655), 68 per cent more fresh water (from 2.5 million cubic meters to 4.2), and 85 per cent more electricity (from 206 Megawatts per hour to 380). The copper produced by the Riotinto mine is presented as a necessary raw material of the energy transition, though such a rise of inputs necessary to production unveils the 'fossilisation of renewable energy' at stake behind green growth narratives (Raman 2013). This expansion process materializes as infrastructure that combines promises of innovation and socio-technical path-dependency: 'infrastructures are seldom built by system builders from scratch. They are instead brought into being through compromised, improved projects of maintenance and repair' (Anand, Gupta and Appel 2018: 12).

The Limits of Novelty

In a pioneer article on the ethnography of infrastructure, Susan Star underlines embeddedness as one of the key features, in the sense that infrastructure 'is sunk into and inside other structures, social arrangements, and technologies' (Star 1999: 381). Mining recovery scenarios, where new projects develop in historical minescapes (Ey and Sherval 2016), highlight this embeddedness particularly clearly.

Despite Atalaya Mining officials' repeated claims of rupture and modernity, the frontier between Riotinto old and new mine is blurry. Mining past is present everywhere in mining infrastructure. 'Everything you see as old and oxidised is from the previous mine', explained Maria as she drove us through tubes, rails and mills that crossed the processing plant, mingling with modern installations she described as 'shiny and new': it was part of the region's cultural heritage, and they could not remove it. But many old installations have been rehabilitated for contemporary production. For instance, the three tailings ponds, which extend over 580 ha, were built up to seventy years ago. According to Felix, the project manager, it is a guarantee of the solidity of the basins – one of the most contentious aspects of the mine, since in case of rupture, it would be an environmental disaster, such as the one that occurred in 1998 in Aznalcóllar, another Andalusian mine. The expansion project presented in 2020 involves enhancing the capacity of these tailings ponds to host 161 million tons of additional waste by raising their walls up to 42 meters above sea level. The promise of solidity is based on the long-standing nature of the facilities and on their previous administrative authorization: being an 'authorized depot that meets the highest safety standards', the best alternative would be to 'give continuity to the current method of operation'.[17] Therefore, beyond their innovative stance, mining technological promises are 'embedded' (Star 1999) in an infrastructural continuity that foreshadows several kinds of potential resistance.

Environmental organizations endlessly warn about the risks surrounding the accumulation of mining waste in the tailings ponds of the Riotinto mine. They accuse the regional authorities who grant their authorization of 'playing Russian roulette' with the threat of rupture of these ponds, which are full of 'toxic mud'.[18] However, in this historical mining basin, the ecological critique has nearly no resonance. In this mining area, whose industrial past infuses daily life through ruins and memory, and where young generations face endemic unemployment, dwellers mainly worry about the creation of jobs for themselves or their fellows. Therefore, environmental expertise that contradicts corporate expertise on mining hazards does not put at risk extractive projects as much as it might in other contexts (F. Li 2015). As weak as is environmentalism in the area, the Riotinto project still faces two major vulnerabilities. The first one relates to local social history. Between the 1960s and 2000s, miners organized themselves into very powerful unions that went as far as running the mine by themselves, through a cooperative, at the end of the twentieth century (Gallego Pajares and Ruiz Ballesteros 2009). Atalaya Mining thus faces the challenge of protecting the mining infrastructure from the workers' well-known capacity for strikes and blockades. The second threat concerns the uncertain profitability of an activity that is subject to market fluctuations, geological contingencies specific to the depleted deposit exploited by Atalaya and other unpredictable events such as a pandemic. Faced with these vulnerabilities, Atalaya Mining has adopted the same solution: subcontracting jobs, which ensures lower production costs and flexibility while limiting workers' protest capacity.

Mining is divided into two main branches: ore extraction and metallurgical processing. For the Riotinto project, Atalaya Mining has a payroll of around 450 people, most of whom work in metallurgical installations. They are clearly separated from the extraction area by a national road that runs between the processing plant and the open pit. Ore extraction is subcontracted to two companies: one carries out the daily blasting of the rock, while the other transports the extracted rock to the tailings pond or to the processing plant according to Atalaya geologists' instructions. Each of these companies subcontracts a series of tasks to a dozen smaller enterprises. The production regime of the Riotinto mine is thus based on a cascading pyramid of subcontractors, dominated by Atalaya with a base of workers with multiple statuses (permanent or fixed-term employees, temporary workers, trainees) spread across dozens of subcontractors that employ close to 500 people. As observed elsewhere (Rubbers 2020), this subcontracting system leads to the segmentation of the labour force, leading to unequal rights. As most of the subcontractors are registered in the construction or transportation sector, they are not subjected to the mining collective agreement, which includes measures to prevent and compensate for occupational health risks specific to the sector. Delegation of work to contractors also enables the Atalaya company to 'subcontract risks' (Thébaud-Mony 2008). Subcontracted workers who operate in the pit are daily exposed to silica-laden dust, rockfall, accidental explosions and other risks associated with the operation of large machines.

While these risks are known to workers, subcontracting presents many obstacles to collective action. As the workforce is divided into a multitude of teams, workplaces and time systems, collective spaces and moments are limited. The multiplication of subcontractors also favours workers' geographical dispersion. Whereas in the nineteenth and twentieth centuries most of the workforce came from and resided in the municipalities of the area, today it is largely composed of people who live more than an hour's drive away. This workforce dispersion hinders informal sociability outside the workplace. These were central to the formation of a collective of unionized miners, strongly politicized and anchored in the space of the mining basin in the twentieth century (Ruiz Ballesteros 1998). These obstacles to collective action are strengthened by the precariousness of the working conditions of a predominantly contractual workforce. Uncertainty about the renewal of their contract weighs on the willingness of workers to speak out or to set up a union branch. In these conditions, protest unionism is rare. Works councils are gradually being overtaken by delegates of the Workers' Trade Union (USO). Two members of this notoriously moderate union claimed to have a conciliatory approach as opposed to the 'warlike style' of traditional local unions.[19]

To sum up, technological continuity more than radical change characterizes the Riotinto project. In this complex entanglement of mining past, present recovery and future projections (D'Angelo and Pijpers 2018), the typical corporate narratives that 'play up the benefits of "new" mining over the "old" mines past' (F. Li 2015: 14) conceal 'modern' infrastructure's embeddedness in local geology and society. The main shift lies in the labour organization implemented by Atalaya Mining: faced with uncertain profitability, shareholders and officials have adopted a subcontracting system that breaks with the history of mining paternalism and union organization (Ruiz Ballesteros 1998).

Conclusion

Discourses on the necessity of 'sustainable mining' to overcome a multilayered crisis rely on temporal narratives that work as a justification scheme. But they also produce tangible effects on politico-economic activities. Industrial encounters and public-private partnerships celebrated and strengthened in 'theatres of optimism' are crucial to the development of mining revival programmes. These fictional anticipations perform 'green' extractivism, in the theatrical sense of a collective representation targeting a public (the justification process), but also following the idea that discursive acts are performative, insomuch as they can shape the materiality of mining. Regardless of if they are realistic or not, promises of 'sustainable mining' produce expectations and trigger activities in people's lives (Luning 2012b). But while impoverished members of historical mining societies expect copper extraction recovery to boost industrial employment for themselves or their relatives, mining materiality and corporate strategies to ensure profitability lead to growing work precariousness.

Although the latter hinders miners' collective action, the growing disappointment of workers and inhabitants might feed a will to protest unfulfilled industrial promises. It is more likely that technological promises might be challenged by the geology of the actual mine, which has been durably shaped by extraction and may resist expansion projects, leading to accidents, activity interruption and even major ecological damages. Indeed, infrastructures are deeply embedded in socio-technical realities that at least partially escape designers' intentions. Projects take place within 'minescapes' (Ey and Sherval 2016) whose specific socio-technical features complicate technological promises.

In light of these (potential) complications of the promises of 'green' extractivism and their materializations in Andalusia, I make some final comments below regarding the connection between the call for 'energy transition' and renewed forms of exploitation of human and more-than-human nature. Performances of new extractivism rely on a complex set of temporal narratives where technological innovation mingles with the industrial past. Mining temporalities are not linear (D'Angelo and Pijpers 2018): the past, present and future of mining entangle in different ways, in line with perspectives and interests that might diverge. Industrial and political promoters of the mining revival celebrate Andalusia's mining history as belonging to museums and cultural heritage, while reopened mines materialize promises of sustainable extraction. But the revival of mining builds upon historical modes of exploitation that reproduce the social and environmental contradictions of 'fossil capital' (Malm 2016). The materializations of 'green' extractivism reveal how the 'energy transition', as long as it remains confined within the framework of the dominant socio-economic order that it aims to preserve, repeats the failures of the Anthropocene.

Doris Buu-Sao is an Associate Professor of Political Science at the University of Lille. Her work concerns the extractive industries, from the policies that promote them to the territories where they develop, from experiences of work to that of living near polluting installations. Before studying the European mining revival in Andalusia, she investigated oil extraction in the Peruvian Amazon, from which came the book *Le capitalisme au village: pétrole, Etat et luttes environnementales en Amazonie* (Capitalism in the Village: Oil, the State and Environmental Struggles in Amazonia, CNRS éditions, 2023).

Notes

1. Communication from the Commission to the European Parliament and the Council: 'The Raw Material Initiative – Meeting our Critical Needs for Growth and Jobs in Europe', 11/04/2008, p. 10.
2. Interview with the technical advisor of the Junta de Andalucia's General Secretary of Industry, Energy and Mining. October 2019, Sevilla.
3. Data from an ongoing ethnography initiated in 2018. Beyond the observation of the public encounters described in the following section, and interviews with public and private actors of mining recovery in Sevilla, the fieldwork focuses on the Riotinto mining area, mainly in Nerva and Riotinto villages (see Figure 2.1), where I conducted observations and interviews with mining enterprise officials and workers, as well as with local dwellers, activists and public authorities.
4. For instance, it is one of the four mines that are part of the INFACT (Innovative, Non-Invasive and Fully Acceptable Exploration Technologies) European project: https://www.infactproject.eu/.
5. Asterisks indicate quotes in English in the fieldwork context. All other quotes are translated from Spanish.
6. According to the Spanish delegate at the 'environment and raw materials' European programme committee, during her conference on 'Research and innovation on raw materials in Horizon Europe'.
7. See 'Raw Materials', https://eit.europa.eu/our-communities/eit-rawmaterials, retrieved 14 January 2022. EIT RawMaterials has been initiated and is funded by the European Institute of Innovation and Technology (EIT), a body of the European Union.
8. According to the Vice-President of the CSIC Foreign affairs division.
9. Interview with an officer of the Junta de Andalucía's mining department, Sevilla, October 2019.
10. See 'Sustainable Mining', MMH website, https://mmhseville.com/mineria-sostenible/, retrieved 14 January 2022.
11. OECD, 'Mining Regions and Cities: Case of Andalusia, Spain', 2021, OECD Rural Studies, Paris, p. 159.
12. Interview with the officer of the Andalusian mining department.
13. Atalaya Mining, 'Documento 1: Proyecto técnico ambiental', *Solicitud de modificación sustancial de la AAU del proyecto Riotinto con motivo de la actualización del proyecto minero vigente*, August 2020, p. 23.
14. Interview with an Atalaya Mining engineer, November 2021.

15. 'Atalaya encarga la construcción de una planta para la producción de metal puro en Riotinto', *Tinto Noticias*, 01/10/2022, https://tintonoticias.com/atalaya-encarga-la-construccion-de-una-planta-para-la-produccion-de-metal-puro-en-riotinto/, retrieved 14 January 2022.
16. Atalaya Mining, 'Documento 1: Proyecto técnico ambiental', *Solicitud de modificación sustancial de la AAU del proyecto Riotinto con motivo de la actualización del proyecto minero vigente*, August 2020.
17. Atalaya Mining, ibid.: p. 144.
18. Ecologistas en Acción (Federación Andaluza), 'La Junta juega a la ruleta rusa con las balsas de lodos tóxicos de Atalaya Riotinto Minera', 2 September 2021, https://www.ecologistasenaccion.org/178381/propuesta-contraria-de-la-junta-a-las-balsas-de-lodos-toxicos-de-atalaya-riotinto-minera/, retrieved January 14 2022.
19. Interview with the head of the USO Andalusia industrial federation, Seville, October 2021; interview with a drill worker and USO shop steward at a local mine, Valverde del Camino, November 2021.

References

Alonso, Steven, Judith Clifton and Daniel Díaz-Fuentes. 2016. 'The Slow Pace of Public Administration Reform in Spain', in Gerhard Hammerschmid et al. (eds), *Public Administration Reforms in Europe: The View from the Top*. Cheltenham: Edward Elgar Publishing, pp. 205–18.

Anand, Nikhil, Akhil Gupta and Hannah Appel. 2018. 'Introduction: Temporality, Politics, and the Promise of Infrastructure', in Nikhil Anand, Akhil Gupta and Hannah Appel (eds), *The Promise of Infrastructure*. Durham, NC: Duke University Press, pp. 1–38.

Andreucci, Diego, and Christos Zografos. 2022. 'Between Improvement and Sacrifice: Othering and the (Bio)Political Ecology of Climate Change', *Political Geography* 92: 1–11.

Arenas Posadas, Carlos. 2016. *Poder, Economía y Sociedad en el Sur: Historia e Instituciones del Capitalismo Andaluz*. Sevilla: Centro de estudios andaluces.

———. 2017. 'Riotinto: el Declive de un Mito Minero (1954–2003)', *Revista de historia industrial* 26(69): 109–42.

Aspers, Patrik, and Asaf Darr. 2011. 'Trade Shows and the Creation of Market and Industry', *The Sociological Review* 59(4): 758–78.

Beckert, Jens. 2013. 'Imagined Futures: Fictional Expectations in the Economy', *Theory and Society* 42(3): 219–40.

Bina, Olivia. 2013. 'The Green Economy and Sustainable Development: An Uneasy Balance?', *Environment and Planning C: Government and Policy* 31(6): 1023–47.

Boyer, Dominic. 2018. 'Infrastructure, Potential Energy, Revolution', in Nikhil Anand, Akhil Gupta and Hannah Appel (eds), *The Promise of Infrastructure*. Durham, NC: Duke University Press, pp. 223–43.

Brown, Nik, Brian Rappert and Andrew Webster. 2017. 'Introducing Contested Futures: From Looking into the Future to Looking at the Future', in Nik Brown, Brian Rappert and Andrew Webster (eds), *Contested Futures: A Sociology of Prospective Techno-Science*. London: Routledge, pp. 3–20.

Buu-Sao, Doris. 2021a. 'Extractive Governmentality at Work: Native Appropriations of Oil Labor in the Amazon', *International Political Sociology* 15(1): 63–82.

———. 2021b. 'Faire Advenir La "Mine Durable"? Discours Institutionnels et Impératif de Relance Minière, de l'Union Européenne à l'Andalousie', *Gouvernance* 18(2): 16–41.

Chastagnaret, Gérard. 2000. *L'Espagne, Puissance Minière dans l'Europe Du XIXe Siècle*. Madrid: Casa de Velázquez.

______. 2017. *De Fumées et de Sang: Pollution Minière et Massacre de Masse. Andalousie – XIXe Siècle*. Madrid: Casa de Velázquez.
D'Angelo, Lorenzo, and Robert J. Pijpers. 2018. 'Mining Temporalities: An Overview', *The Extractive Industries and Society* 5(2): 215–22.
Death, Carl. 2010. *Governing Sustainable Development: Partnerships, Protests and Power at the World Summit*. London: Routledge.
Del Mármol, Camila, and Ismael Vaccaro. 2020. 'New Extractivism in European Rural Areas: How Twenty-First Century Mining Returned to Disturb the Rural Transition', *Geoforum* 116: 42–49.
Desage, Fabien, and Jérôme Godard. 2005. 'Désenchantement Idéologique et Réenchantement Mythique des Politiques Locales', *Revue française de science politique* 55(4): 633–61.
Dunlap, Alexander, and Jostein Jakobsen. 2020. *The Violent Technologies of Extraction: Political Ecology, Critical Agrarian Studies and the Capitalist Worldeater*. London: Palgrave Macmillan.
Ey, Melina, and Meg Sherval. 2016. 'Exploring the Minescape: Engaging with the Complexity of the Extractive Sector', *Area* 48(2): 176–82.
Franquesa, Jaume. 2018. *Power Struggles: Dignity, Value, and the Renewable Energy Frontier in Spain*. Bloomington: Indiana University Press.
Gallego Pajares, Ricardo, and Esteban Ruiz Ballesteros. 2009. 'Minas de Río Tinto Sociedad Anónima Laboral: Un Modelo Sociolaboral Alternativo para la Reactivación de la Minería Onubense', *Trabajo: Revista Andaluza de Relaciones Laborales* 1: 35–45.
García Fernández, Javier. 2019. *Descolonizar Europa: Ensayos Para Pensar Históricamente Desde El Sur*. Madrid: Brumaria.
Goldstein, Jesse. 2018. *Planetary Improvement: Cleantech Entrepreneurship and the Contradictions of Green Capitalism*. Cambridge, MA: MIT Press.
Goldstein, Jesse, and David Tyfield. 2018. 'Green Keynesianism: Bringing the Entrepreneurial State Back in(to Question)?' *Science as Culture* 27(1): 74–97.
Gudynas, Eduardo. 2019. 'Development and Nature: Modes of Appropriation and Latin American Extractivisms', in Julie Cupples, Marcela Palomino-Schalscha and Manuel Prieto (eds), *The Routledge Handbook of Latin American Development*. London and New York: Routledge, pp. 389–99.
Hernández Ramírez, Macarena, and Esteban Ruiz Ballesteros. 2007. 'Identity and Community: Reflections on the Development of Mining Heritage Tourism in Southern Spain', *Tourism Management* 28(3): 677–87.
Horowitz, Leah S. 2011. 'Interpreting Industry's Impacts: Micropolitical Ecologies of Divergent Community Responses', *Development and Change* 42(6): 1379–91.
Jasanoff, Sheila, and Sang-Hyun Kim. 2015. *Dreamscapes of Modernity. Sociotechnical Imaginaries and the Fabrication of Power*. Chicago: University of Chicago Press.
Joly, Pierre-Benoît. 2010. 'On the Economics of Techno-Scientific Promises', in Madeleine Akrich et al. (eds), *Débordements: Mélanges Offerts à Michel Callon*. Paris: Presses des Mines, pp. 203–22.
Kirsch, Stuart. 2010. 'Sustainable Mining', *Dialectical Anthropology* 34(1): 87–93.
Larkin, Brian. 2013. 'The Politics and Poetics of Infrastructure', *Annual Review of Anthropology* 42(1): 327–43.
Laurent, Brice, and Julien Merlin. 2021. 'Investissement, anticipation, planification: la politique temporelle des activités extractives', *Revue Gouvernance/Governance Review* 18(2): 42–62.
Leivestad, Hege Hoyer, and Anette Nyqvist. 2017. 'Individuals and Industries: Large-Scale Professional Gatherings as Ethnographic Fields', in Hege Hoyer Leivestad and Anette Nyqvist (eds), *Ethnographies of Conferences and Trade Fairs: Shaping Industries, Creating Professionals*. New York: Springer, pp. 1–21.

Li, Fabiana. 2015. *Unearthing Conflict: Corporate Mining, Activism, and Expertise in Peru*. Durham, NC and London: Duke University Press.

Li, Tania Murray. 2007. *The Will to Improve: Governmentality, Development, and Practice of Politics*. Durham, NC and London: Duke University Press.

Luning, Sabine. 2012a. 'Corporate Social Responsibility (CSR) for Exploration: Consultants, Companies and Communities in Processes of Engagements', *Resources Policy* 37(2): 205–11.

———. 2012b. 'Processing Promises of Gold: A Minefield of Company–Community Relations in Burkina Faso', *Africa Today* 58(3): 23–39.

———. 2018. 'Mining Temporalities: Future Perspectives', *The Extractive Industries and Society* 5(2): 281–86.

Malm, Andreas. 2016. *Fossil Capital: The Rise of Steam Power and the Roots of Global Warming*. London: Verso.

Marín Arce, José María. 2006. 'La Fase Dura de la Reconversión Industrial: 1983–1986', *Historia del Presente* 8: 61–101.

Martínez-Alier, Joan. 2002. *The Environmentalism of the Poor: A Study of Ecological Conflicts and Valuation*. Cheltenham: Edward Elgar Publishing.

Mitchell, Timothy. 1999. 'Society, Economy, and the State Effect', in George Steinmetz (ed.), *State/Culture: State-Formation after the Cultural Turn*. New York: Cornell University Press, pp. 76–97.

Mosse, David. 2005. *Cultivating Development: An Ethnography of Aid Policy and Practice*. London: Pluto Press.

Phadke, Roopali. 2018. 'Green Energy Futures: Responsible Mining on Minnesota's Iron Range', *Energy Research & Social Science* 35: 163–73.

Rajak, Dinah. 2011. 'Theatres of Virtue: Collaboration, Consensus, and the Social Life of Corporate Social Responsibility', *Focaal* 60: 9–20.

Raman, Sujatha. 2013. 'Fossilising Renewable Energies', *Science as Culture* 22(2): 172–80.

Rubbers, Benjamin. 2020. 'Mining Boom, Labour Market Segmentation and Social Inequality in the Congolese Copperbelt', *Development and Change* 51(66): 1555–78.

Ruiz Ballesteros, Esteban. 1998. *Minería y Poder: Antropología Política en Riotinto*. Huelva: Diputación provincial de Huelva.

Star, Susan Leigh. 1999. 'The Ethnography of Infrastructure', *American Behavioral Scientist* 43(3): 377–91.

Thébaud-Mony, Annie. 2008. 'Sous-Traitance des Risques', in Annie Thébaud-Mony (ed.), *Travailler Peut Nuire Gravement à Votre Santé: Sous-Traitance Des Risques, Mise en Danger d'autrui, Atteintes à la Dignité, Violences Physiques et Morales, Cancers Professionnels*. Paris: La Découverte, pp. 83–123.

Van Lente, Harro. 2017. 'Forceful Futures: From Promise to Requirement', in Nik Brown, Brian Rappert and Andrew Webster (eds), *Contested Futures: A Sociology of Prospective Techno-Science*. London: Routledge, pp. 43–64.

Van Teijlingen, Karolien. 2016. 'The "Will to Improve" at the Mining Frontier: Neo-Extractivism, Development and Governmentality in the Ecuadorian Amazon', *The Extractive Industries and Society* 3(4): 902–11.

— Chapter 3 —

Unearthing the Buried Past of Brazil's Former Gold Mines

Andreza Aruska de Souza Santos

In this chapter, I bring the city of Ouro Preto to the forefront to discuss the state-led formation of Brazil's past and local forms of resistance to it. Ouro Preto is a place that tells a positive history of Brazil like no other. A major gold exporter from the Americas to Europe during colonial times, Ouro Preto combines architectural legacy – emphasizing the creativity of local artisans – with anti-colonial struggle, which were both pivotal for Brazil in its nation-building momentum in the early twentieth century. Memories of independence were stressed by the Brazilian state through monuments and street-naming in Ouro Preto, and at the same time, other stories were silenced. State-led heritage policies did not underline that plots against Portugal in Ouro Preto were mainly led by a Brazilian-born elite with non-explicit anti-slavery ideals. Slavery, instead, was filtered in the process of creating a positive image of Brazil that is racially democratic and nonviolent. This chapter will discuss the hegemonic narrative of Brazil's past in its search for a consensual history as well as resistance to it, when *Ouropretanos* bring conflict to the centre of history. I look at gold mines in contemporary Ouro Preto, where the spirits of slaves find a home and where tour guides' dialogue with these spirits to reveal the cruelty of slavery, a much-hidden aspect of Brazil's history.

Post-colonial states had to decide on new imaginaries for the nation (Cardina and Nascimento Rodrigues 2020). Hegemonic narratives were commonly related to struggles for independence, even in places where there was no independence war (Gonlugur and Sezer 2002). All too often, however, nation-building narratives with a focus on independence movements minimized colonial history. This is the case in Brazil. When I discuss hegemonic narratives in this country, 'forgetting' is a key aspect of it. Politically, dealing with the past needed to both create a 'positive self-image' for Brazil and allow for a new future of reconciliation in a racially, regionally and economically divided society. My writing is thus situated in an interdisciplinary discussion that looks at politics and nation-building; economic, political and international transitions that acknowledge competing pasts; as well as an anthropological analysis that intersects different ways of remembering in former gold mines.

A 'divine memory', as I will explore below when discussing mine tours, is not new to the intersection of memory and anthropological studies. Ritual memories, such as cosmologies of witchcraft and phantasmagorical rumours, can challenge written and material ways of remembering (Shaw 2002). Such memories are thus important in contexts where written and material reckonings have silenced some aspects of history that can resurface in religious rituals. The association between colonial slavery and spirits is thus common (Hale 1997; Routon 2008; Stoller 1997), as it is in other violent pasts that are often retold through narratives involving spirits and ghosts. The legacy of the Gulags in Siberia is retold with an emphasis on ghosts that inhabit buildings (Ulturgasheva 2017). Memories of apartheid in South Africa include expressions of thirst in rituals that remember oppression (Comaroff 1985). This chapter adds to a body of literature that looks at spiritual possession to identify 'ritually stylised forms of social mimesis' (Routon 2008). Through spiritual possession, colonial memories gain a phenomenological dimension (Stoller 1997). Spiritual manifestation in Ouro Preto thus fosters a 'living past' (Harris 2007).

In the case of Ouro Preto's mine tours, even though visitors do not witness moments of spiritual possession, the history of slavery there is expressed in detailed accounts about spirits that inhabit the mines. Spirits can affect the living, and visitors have reported getting headaches as a result. In addition, earlier working conditions – including sounds, sights and sensations – have been re-created in the mines.

In what follows, I first look at tangible forms of Brazil's past and how colonial towns such as Ouro Preto were represented in laws, decrees and preserved monuments to represent a dominant version of Brazil's history. I show how slavery was forgotten in much of

Brazil's textual and material history. Secondly, I discuss embodied and oral memories of the forgotten slavery in a mining economy. I discuss the difficulty of this past for government authorities and why memories of the past have, quite literally, not yet reached the surface. The forgotten history of Ouro Preto's mine slaves is remembered when former slaves' spirits influence visitors, suggesting a potentially innovative and transformative shift in power dynamics, where in the formerly controlled spaces of oppression the spirits in the mines are never those of former owners but of oppressed slaves. This discussion brings us to the third and final theoretical debate of this chapter, which is related to power. Remembering is subject to power, by which I mean that the interests and capacity of those in the present can dictate what past resurfaces and how and when. I finish by discussing how memories in mines, which are at once intangible and violent and for those reasons transgressive against the hegemonic history of Brazil, are also limited in effect.

When descendants of slaves speak of their ancestors as not mere victims but as present subjects controlling territories, this may be confused with seeing slaves as rulers or agents of their own history. In grappling with a subversive narrative, I also recognize forms of colonial dominance to avoid writing a 'revisionist' history (Shaw 2002: 20). I discuss the limits of counternarratives of Brazil's hegemonic history. In so doing, this chapter engages with other nuanced anthropological writings on Brazil that focus on religious temples and music festivals as places where black Brazilians are able to control their narrative. Similar to this study of former gold mines, what much of that literature emphasizes is space-time limits, and counter-hegemonic memories may be restricted to some temples and/or festivals (Baiocchi and Corrado 2010; Perry 2016; Sheriff 1999).

Methodologically, this chapter is based on archival research and ethnography. I lived in Ouro Preto in 2013 during my doctoral fieldwork and visited the city annually from 2016 to 2019 (before the Covid-19 pandemic). During these periods, I visited the city's local archives and Rio de Janeiro's (the former capital of Brazil) and Brasilia's (the current capital of Brazil) archives to collect material (such as official government magazines that justify heritage site choices) on the process of heritage-making in Brazil, studying which memories were selected to form museums, monuments and preserved sites. Using ethnography, I have observed how people have resisted or conceded to the past invented for them and chosen the elements that they have insisted on enacting.

Brazil Needed a Past

Countries with colonial legacies such as Brazil cannot ignore the burden of their slavery past, which manifests as poverty, racial discrimination and inequality. And yet, in Brazil, government-led narratives have focused on racial mixture, the so-called *mestizaje*, as have most other Latin American countries (Telles and Paschel 2014). In practice, what *mestizaje* means is that Brazil has glossed over its history of violence, forced migration and oppression in slavery with a nation-building narrative for a multiethnic society. Heritage policies are also influenced by this nation-building narrative. Monuments and the preservation of cityscapes show how Brazilian-born groups gained focus in the fight for independence. In addition, monuments also tell the story of the creativity of local artists, often from a mixed race background, who built European architecture in Brazilian coastal or rural lands. This tangible history of Brazil is aimed at uniting a Brazilian-born elite, slaves and indigenous groups in a general duality of 'Brazilians' versus 'the Portuguese'. In addition, creativity and racial mixing emphasized by colonial artists bring to the fore ideals of meritocracy and social ascension (De Souza Santos 2019a).

Heritage policies gained special momentum during Getúlio Vargas' administration (1930–45). Governing after regional elites from Minas Gerais and São Paulo had taken turns holding Brazil's presidency for thirty-two years (1898–1930), Vargas hoped to unite Brazil, and to do so he put forward a cultural agenda that gained traction with its focus on a common enemy – colonizers – in cultural heritage-making. Former colonial cities were re-signified as spaces where the disputes between Brazilians and colonizers were highlighted.

Large urban areas, entire city centres, or specific buildings in cities such as Ouro Preto, Congonhas, Tiradentes, Olinda and Salvador gained a monumental status as they told the story of a colonial Brazil that resisted invasion. The 'editors of history', the public servants of SPHAN (National Historic and Artistic Heritage Institute) in charge of selecting and historicizing monuments in Brazil, retrospectively assigned meaning to the past. Looking for signs of 'Brazilianess', these editors emphasized the use of new material to highlight the creativity of local artists, and stories of resistance against colonial power, which were told through the preservation of cityscapes – old prisons, palaces, houses and places of conspiracy. With a growing emphasis on independence movements, Ouro Preto became the most important of all Brazilian heritage sites, as it was the home of the Minas Gerais Conspiracy (1789) (Maxwell 1973).

The Minas Gerais Conspiracy is one of the most documented colonial rebellions in Brazil, and Ouro Preto, where the plan to rebel-was conceived, became the Vargas government's pride and joy. The conspirators were part of a local elite who saw in the collection of taxes an opportunity to rebel against the colonial power and seek independence. Perceiving the growth in taxes imposed by Portugal in a shrinking economy as a shared problem, the group planned to rise up in rebellion during a tax collection exercise. However, the same economic interests that motivated the conspiracy led to betrayal within: the plot was exposed before it could take place. This could have been an unimportant anti-colonial conspiracy, and yet it became the most important one. When the conspirators were arrested, the least privileged of them assumed full responsibility and was the only one charged with the death penalty. His head displayed in a public square, which is today named after him, Tiradentes was the hero of that movement. The making of Tiradentes as a national hero overshadowed other anti-colonial rebellions that had a stronger message about inequality and slavery. Not only that, but the focus on the Minas Gerais conspiracy also played down slavery in Ouro Preto. At the time (in the eighteenth century) up to 80 per cent of the population of Ouro Preto consisted of slaves (Maxwell 1973). In 1933, this city of environmental and human destruction was, however, preserved as a place where Brazilians resisted Portuguese rule (Decree 22928, passed on 1933) (De Souza Santos and Iamamoto 2019).

The history of a small group of Brazilian-born elites fighting against tax payments to the Portuguese Crown in a declining gold economy was historicized as Brazilians plotting to throw off the Portuguese rule. However, the lack of commitment to the abolishment of slavery by the affluent rebellious group is not commonly addressed when the Minas Gerais Conspiracy is discussed in official government reports (IPHAN n.d.), and the history of slavery lingers in the background while monuments and streets are named after the Minas Gerais conspirators and the remains of their bodies and personal objects populate museums and mausoleums in the town.

Ouro Preto is thus a good place to discuss the national sentiment in the Vargas administration. The diminished role of racial, regional and economic inequalities is materialized in a town that celebrates Brazil's struggle for independence without attention to slavery or indigenous history. The results of Vargas' policies, which had lasting effects in Brazil, as I will discuss, are ambiguous. While a broad concept of inclusionary citizenship emerged in Brazil, the history of racism and slavery were kicked into the long grass. Brazil's 'difficult heritage' (Macdonald 2009) of racial and economic inequality is thus a threat to the national ideal of social cohesion based on the idea of racial democracy and meritocracy, and slavery heritage has long been avoided by the architects of memory in Brazil (Chuva 2009).

Despite the hegemonic government-propagated memory, heritage sites have become sites of contestation and pushback regarding such top-down narratives of a socially cohesive Brazil. Immaterial heritage, which started especially after the 1980s in Brazil and elsewhere (an initiative led both by national governments and UNESCO), is a form of inclusion for excluded groups, especially indigenous and black Brazilians (Hatzikidi 2020). Intangible heritage would allow these groups to be included on heritage sites' lists. In places such as Ouro Preto, however, the predominance of tangible heritage, with colonial churches, squares of colonial power and old colonial houses all well preserved, has meant that the intangible aspects of history are often overshadowed, and inclusion through cultural heritage has often been delayed in these towns. Slavery remains essentially neglected in Ouro Preto's heritage sites.

Despite the fact that Ouro Preto's colonial architecture is unquestionably the result of slavery violence, hunger and wars in the quest for gold in Brazil, there is a separation between the mines beneath the surface, which are invisible to the eye, and the products of the gold, which are the colonial houses and churches above in an acclaimed baroque architecture. This separation further cements the invisibility of the history of slavery in town. The same separation also exists when looking at urban sprawl. Whereas slaves in the past worked in the mines, the poor residents of Ouro Preto today, mostly dark-skinned, occupy the peripheric areas of the city, away from the historic centre. Some of these same residents who are spatially and culturally excluded from the city are the tour guides of the gold mines, and before describing their work, I will explain some of the socio-spatial dynamics in town.

Ouro Preto's Socio-spatial Divide and the Peripheral Place of Tour Guides

In the eighteenth century, when Ouro Preto became a major gold exporter from the Americas to Europe, colonial ruling and wealthy families changed the hilly landscape of the town with mansions and temples to showcase their power and control through architectural design (Rama and Chasteen 1996). The colonial architecture has resisted the pressures of time, amongst other reasons because of the violent population decline that followed. The gold in town did not last more than fifty years of exploration. The population of the town declined in the second half of the eighteenth century, but because the city remained the capital of the state of Minas Gerais, and politics was its new gem, mass out-migration followed only after 1897, when Ouro Preto relinquished its capital status to the newly built city of Belo Horizonte (Batista da Costa 2011). The demographic emptiness of the city meant that the wealth of the golden years carved in soapstone sculptures, baroque churches and colonial mansions was preserved, and the entire city looked like an open museum without visitors. The preservation of the city as a national heritage site in 1933 was aimed at creating a shared collective memory of Brazil as well as enabling the reoccupation of the city. However, reoccupation was not easy. The withdrawal of government functions and the unwelcoming geology for large businesses of a city carved in the mountains delayed repopulation, which only began in earnest in the 1950s, after new mining companies moved to town, attracting new residents, and when a university was inaugurated, inviting students and more tourists to the area (De Souza Santos 2019a). Because mining is no longer as labour-intensive as in colonial times (De Souza Santos 2019b) and it happens mainly on the outskirts of Ouro Preto's vast district, most of the local population in Ouro Preto have found jobs in tourism, public administration or in services at the local university. This has led to the stigma of the city as catering for 'others' outside of the town and paying its workforce the minimum wage.

At first, tourism was seen by locals as an engine to promote jobs in times of heritage making. However, the local population eventually realized that such jobs were occasional and poorly paid, and that the city was no longer 'theirs'. Additionally, the local university has mainly received students and faculty from other Brazilian towns, as the majority of local residents do not have the educational background to become the educated elite of tomorrow. Education and tourism, the vectors of Ouro Preto's resurgence on the national scene, are thus locally seen as segregating elements (De Souza Santos 2019a).

Because the majority of residents dwell on the slopes of the hills while the preserved centre caters mainly for students and tourists coming from elsewhere, locals commonly refer to visitors as 'colonizers'. The newcomers take the best of what the city offers (centrally preserved houses and educational and cultural opportunities) for a limited amount of time and then leave, while the majority of residents live in fear of eviction from their houses, which are perceived as an aesthetic nuisance.

Those same residents who are excluded from central housing and from the statues and monuments of the city are black and poor and work as tour guides. As such they have a unique interpretative role in the city (Reed 2002). In lending their voices to explain the city's past and present, the ambiguity of their lives often takes centre stage. The interpretation and imagination offered on city walks represent contradictions related to the reality of the precariousness of local guides' lives versus the wealth of touristic places. Tour guides in Ouro Preto are thus caught between privilege and exclusion.

While guides endeavour to satisfy vacationers by offering a positive image of the city, as residents, they are often excluded from the city's cultural events and live outside the main touristic, commercial and cultural zone. The ambiguity of their lives in a city where residents have maintained a narrative about social and spatial divides since colonial times is overwhelmingly apparent in former gold mines. Mines are the seed that created the city's wealth. However, those spaces are beneath the ground and invisible. Using mines to unearth the contrasts of contemporary Ouro Preto is fitting, it being a town where the workforce lives beyond the visible heritage skylines (under the earth in mines or on top of hills nowadays) and where the gems of the city are supplied by those who cannot make use of them (gold in the past, or tourism and education in the present). The display of past and present features is not limited to these examples, and colonial violence also resurfaces in tours, as I narrate in the next section.

Remembering Slavery in Mines

In this section, I discuss mines as spaces of colonial violence, with an emphasis on the spirits of slaves that inhabit them. Ghostlike memories of slavery illustrate that slaves are found in narratives about Ouro Preto through mystical oral tales rather than local statues, statistics or biographies. Off-the-record spiritual or imagined remembrances by residents enact a history overlooked by official accounts. There are limits to such off-the-record narratives, but I will discuss how they seek to unsettle the hegemonic narrative of 'therapeutic forgetting' (Gonlugur and Sezer 2002) (playing down a slavery past) by exposing underlying tensions. The main material for this discussion is ethnographic, with data collected in 2013.[1]

Ghost Stories

> The majority of the residents are afraid [of mines] because there are many stories of ghosts in town. When slavery finished, this place [the mines] was abandoned – Ouro Preto was abandoned. The few people who remained, the mine owners, began to make up stories of ghosts so that people would become afraid and would not enter the mines, as owners imagined that one day they would be able to extract gold again. But they forgot to say that the stories were not true, and as the years went by, and extraction did not happen again, slavery did not happen again, the stories told became a truth, because every lie told many times becomes true. So, the elderly in Ouro Preto, the majority of them, 98 per cent, are afraid to enter the mines. (Personal interview with a mine tour guide, 2013)

Such widespread fear of the mines led to their abandonment, and later their reoccupation. It is unclear who nowadays owns the mines, but former gold mines in Ouro Preto have for years been both occupied for use as a residence and commercially explored for tourism by slave descendants, who claim ownership.

The stories of ghosts that were shared with me during my fieldwork shifted the narrative from mine owners trying to keep people away to former slaves (i.e. their spirits) selectively allowing entrance. Tourists often find out if they were allowed to enter the mines only after their visits; those who were not welcome may get sick afterwards.

> Mines suck the energy, especially if you have a strong sensibility. It will make you tired, you will yawn, you will start reading a book and after the two first pages you will stop and will have to continue slowly – there are many things. Are you ready to enter? (Mine tour guide, 2013)

Mining settings provoke bodily sensations, and the 'energy' of mine spirits, as tour guides narrate, equally affects the body. This phenomenological experience is different from common forms of gathering knowledge about slavery in Brazil. Slavery in Brazil is usually portrayed in museums, with great emphasis on the violence associated with it – in exhibitions of torture objects, for example – rather than focusing on its cultural or biographic legacies (Hale 1997). This is the case in Ouro Preto. Looking at another colonial context in Brazil, Marcio Goldman draws on his own and others' ethnographic research regarding the region of Ilhéus, Bahia to discuss how the dominant version of the history of the cacao-based economy is based on small estates instead of a large supply of slave labour (Goldman 2013). This dominant version, he clarifies, is far from what documents say about the participation of slaves in the local economy. Collins (2011) looks at Salvador, also in the state of Bahia, and discusses how everyday habits and memories of local black residents in the neighbourhood of Pelourinho have been articulated by local authorities as alienable forms of property. What these combined examples demonstrate is the role of mine tours in moving beyond the common museum displays of torture tools or the lack of (or usurpation of) a narrative concerning slavery's economic and labour contribution that as such denies the existence of elements of past Brazilian society. In the mines, the spiritualization of slavery also allows for previous occurrences to have a phenomenological manifestation. Visitors learn about slavery in the mines both by experiencing the working conditions and when sensing the spirits that are said to reside there. In the following, I share a brief account of a tour through the gold mines of Ouro Preto before engaging further in the analysis of spirits, past and present, in the town:

> November 25, 2013, a drizzly rain persisted all morning before we arrived at the first gold mine. Big J. guided me through the dark, cold and wet mine. While encouraging us to crouch in the narrow and low corridors, he said: 'Imagine if time went back by 200 years and you were a twenty-year-old man, a black man working in mines, poor'. Big J. insisted we should imagine that to let our fears out. As we walked through the mine, Big J. and my city guide stopped walking, and I had to set the pace on my own, for I was the one who was there to 'learn the past through sensations'. I kept on walking; it was very dark, and I could still hear the voices of the two men, though the voices became more distant. 'You need this to understand what the slaves went through back in those days', they said, while I moved through the unknown in a narrow corridor.
>
> The ceiling got lower and lower as I walked. I had to squat in a very uncomfortable and claustrophobic position, and I wondered if I could still move forward or how I would move backwards to leave the place, at which point Big J. yelled: 'Can you imagine if we both went away and you would have to stay here for two or three days?' No, I could not imagine that, and wanted to leave, but Big J. started to hammer against the wall, producing a sharp and repetitive sound. 'Men died from the noise; while digging stones, some particles would reach their eyes making some men blind, while others would have their lungs turned into stone when breathing the powder day after day.' The noise continued for a short while until it became irritating, at which point I tried to make my way back to the exit, but I fell down while moving backwards. It was very slippery. (Field notes, November 25, 2013)

Experiencing humidity, darkness, noise and claustrophobia in the low and narrow corridor in the mine illustrated how Ouro Preto's slaves lived back in the eighteenth century. While stories of violence against slaves are told in mines, there are contradictory attributions of power, as the same people who were enslaved, whipped and chained were also believed to have the power of sorcery (Routon 2008). In Ouro Preto, the power of slaves as sorcerers is indeed common in local stories. But this power is not confined to acts of sorcery in the past, with the spirits of slaves, rather than those of mine owners, dominating the mining territories. Spirits may not allow entrance to the mine, explained the guide of another mine:

> You need to be approved by the spirits of the people there; otherwise you do not enter the mine. There are people who arrive here and ask me to enter, but I arrive at the entrance and I fall down; they [the spirits] beat me up, throw me on the floor, 'with that person you do not enter'. But when the person has a clean soul, a person with the mysticism [*mediunidade*] and the sensibility that matches theirs and mine, we can enter the mine. (Guide in a gold mine, 2013)

The 'people there', the guide explained, are the people who worked as slaves in the mine in the past. When they do not want a sightseer to enter the mine, they beat up the person or the guide by making them fall down or hit their heads against the ceiling or the walls, or by inflicting headaches after a visit. These are seen as the typical signs that the spirits have not granted permission to enter. Local stories involving the spiritual infliction of pain in mines are so common that guides report offering tours for people from across the world and Brazil, while the local population are usually afraid of the mines and do not visit them.

During my tour, Big J. told me a nightmare he had had the previous night. In his report, the dream, like his stories while awake, explained relationships of power and violence during the days of slavery:

> I went to bed, and I had a nightmare as never before. It felt as if my spirit was imprisoned there [in the mine], and there were people there beating me, and I tried to leave but I could not. I felt that affliction that you felt there [when I was left alone in the dark in the mine], and I felt I was going to die of that affliction. Suddenly, when I least expected it, a Preto Velho [old black slave] appeared saying 'you will leave this place and I will help you'. And he helped me to leave, but he was cursing me all the way out. When we got out, I saw a lot of people working inside the mine and my spirit leaving it. … Do you know why I had this dream? Enemy groups from the eighteenth century, enemies there and enemies here, and the blacks did not get along, that is why there was never a rebellion. The majority of the afro-descendants here [of Ouro Preto], they do not like entering the mines. A lot of the time the problem is not you, you are white, it is with another black like me. (Big J., 2013)

Society in Minas Gerais, where Ouro Preto is located, demonstrated great social complexity from the eighteenth century. More than a society of slaves and masters, there was an urban class of artisans, many of them of mixed race (IPHAN n.d.). Likewise, within groups of slaves, as Big J. reported, complexities meant there was often no common ground for relationships. While some slaves would look to make their own personal fortune smuggling gold, others would try to escape slavery in small groups, leaving no space for rebellion. Similar to what Austen notes in the case of the slave trade in Africa, some aspects of the trade are morally problematic to remember (Austen 2001). Of course, the lack of rebellions cannot be blamed on conflicts amongst slaves; the utter violence of slavery did not allow for escape, and for tour guides of the mines, other explanations for the lack of major rebellions by slaves in Ouro Preto – as Big J. noted – can be gained in dreams.

Ghostly mine tours in many respects fill gaps in the official history, as they show where and in what conditions slaves worked and make an explicit connection between the wealth one sees in churches and central constructions in town with the gold that cost human lives. Mine tours also present the biographies of notable black residents, such as the former slave Chico Rei, who according to oral history lived in Ouro Preto at the peak of gold exploration in the first half of the eighteenth century. The story of Chico Rei reflects the limits of the possible power relationship inversion in town as a result of mining. While crouching in the mine where Chico Rei worked, a tour guide called Nelson explained the story of the famous black slave who used to be a king in Africa until he was forced to come to Brazil. In my own words, I summarize the story offered that afternoon:

> Chico arrived in Ouro Preto and worked in an unproductive gold mine, until his master became ill and offered the mine to him. Chico knew about mining and knew that this one had given no gold yet and he would not be able to pay for the mine if there was no gold inside it. Nevertheless, the slave accepted the offer. When he started exploring the mine, mysteriously he finally found gold there. Chico paid for the mine and bought his freedom and that of those who were part of his former tribe. He started a new tribe in Vila Rica and was king once again. He became very rich, and the Crown said 'look, in Vila Rica there is a slave who owns a mine, freed his people, and is today the king of the slaves'. Portugal, afraid of a rebellion, offered Chico Rei all he needed to return to Africa; he could take with him all the wealth he had accumulated. Chico accepted the offer because there he would be free; while in Ouro Preto, though he was king, he was a 'nobody'. Nobody respected him; he was black in the eyes of the whites, and so he never had freedom – he was no longer a slave, but he was always seen as inferior, and so he left. (Nelson, 2013)

Concluding the tour with the story of Chico Rei, who became free and rich yet remained chained to the social organization of the colonial society, suggests that the city's material or immaterial, official or non-registered cultural heritage excludes groups rather than brings them closer together. In the mines, the dualism of oppressed and oppressors gains a new form, when stories of black slaves' spirits express the power of the oppressed. The power of the excluded, however, does not go beyond the mining territories. In the city, there is no statue for black slaves, and the one museum that dedicates a room to slavery exhibits only torture instruments. Additionally, only a few residents visit the mines, partly to avoid the alleged encounters with the spirits and partly because mines, like other tourist attractions, cater mainly to tourists. In addition, the scientific part of history is repressed, when there is no space for a narrative about geological knowledge amongst mining slaves. One is invited only to imagine how mining tunnels were carved and have avoided collapse over the years. Instead, stories of oppression or the invisibility of slavery dominate. Mine tours finish with narratives of enduring separation and privilege:

> We are sad because I think the prefecture or the regional government should grant residents the right to visit the mines, so they would know their history. The majority of the children do not know it; they do not know about the mines. If I were a millionaire, you know what I would like to do? I would like to get some thirty kids, some five or ten from Morro Santana, five from Morro São Sebastião, those black and poor kids from poor neighbourhoods, and would take them to São Jeronimo Museum in Portugal, so they could see what was taken from here. They do not know. We do not know. Have you been to Europe, Nelson? Nelson shook his head. (Conversation between mine guides, 2013)

The Implications of a Selective Politics of Remembering

The idea of collective forgetting to create a shared consensual history – minimizing or erasing traumatic experiences – is a disputed field (Gonlugur and Sezer 2002). Building a future by selectively filtering the past can create a positive self-image. For victims, however, the past is rarely forgettable (Fulbrook 2020). In the case of slavery, racial discrimination and socio-spatial inequality is an everyday struggle for slave descendants in Brazil.

When nation-states silence violent pasts, international solidarity and socio-economic transitions may build momentum to shed new light on silenced memories (Cardina and Nascimento Rodrigues

2020). In this section, I explore how nation-building narratives in Brazil focused on a representation of the past that was selective. Bringing forth a 'Brazilianess' opposed to colonizers, without a focus on inequality amongst Brazilians, perpetuated racial inequality. I will discuss attempts to challenge this narrative in recent policies and grassroots movements. I consider that when nation-building narratives – which are especially important for racial identification – played down racial differences in Brazil, that process was never fully complete. On the one hand, ancestry became less important for racial classification; on the other hand, physical attributes remained important (Telles and Paschel 2014). That is to say that race in Brazil still hierarchizes through skin colour; the mixing of cultures in the making of Brazilian society is not to be confused with equality.

In a brief history of racial theories in contemporary Brazil, what we have is that during the nineteenth century up until the beginning of the twentieth century, just before the Vargas government, the predominant belief was that of 'scientific racism', which claimed the need for migration policies to 'whiten' the population in order to promote modernization and development. Oliveira Vianna, one of the proponents of such a vision, tied social development in Brazil to a gradual elimination of physical and cultural traces of blacks and natives (Nascimento 2003). However, with the sociology of Gilberto Freyre (1963), the emphasis on cultural aspects in the creation of a racial democracy became central. In *Sobrados e Mucambos* (*The Mansions and the Shanties*), first published in 1936, Freyre highlighted the urban changes taking place in Brazil. A social urban class, white-collar and salaried, prevailed in São Paulo and Rio de Janeiro. Considering the growth of cities, the state, industrialization and employment, Freyre emphasized culture and the cultural syncretism – a combination of Africa and Europe – producing a singular Brazilian society, not reduced to its roots (Souza 2000). He presented the *mulato*, the 'type' that had access to cultural encounters and that was capable of rising socially. He valued individual talent and proposed a new social hierarchy, wherein whiteness is associated with cultural attributes. The same rationale that we see in Brazilian political thought also informed the preservation of heritage sites in Brazil, with an emphasis on the creation of a Brazilian society that is mixed in race and can occupy new social spaces through creativity, education and labour.

The echoes of this dominant narrative of a hybrid and harmonious making of Brazilian urban culture are disputed (Sheriff 2001). A dual citizenship for blacks and whites in Brazil is a reality; however, such a duality is framed silently and through a meta-discourse (ibid.). Appearance and not ethnicity, which is what some would call aesthetic racism, is how black and white, dark- and light-skinned Brazilians are differentiated (Souza 2010). Ways of dressing and hairstyles, for example, offer the possibility of altering colour perceptions in Brazilian society. Race is thus often associated with social bonds and intra-class perceptions. For example, a light-skinned Brazilian living in a poor and black-dominated neighbourhood may identify himself as being black – and not to be confused with a dark-skinned Brazilian living in an upper-class, white-dominated neighbourhood. Context is a determining factor. A collective understanding of social location, '*meu lugar*' (my place), is created in the face of a differentiating system (Owensby 1999).

The constant comparison between black and white reveals the basis of racial discrimination. The possibility of shifting social positions adds to the search for a compensating future wherein equality and justice will finally find a home. Such individual possibility of ascension is detached from the cruel realities of everyday life and disregards historical conditions (Ribeiro 2000).

It is precisely this combination of a national narrative of inclusion with a lived experience of racial exclusion that makes racial identification and counter-hegemonic narratives of Brazil's slavery past so difficult in Brazil.

However, if textured racial identities and hegemonic narratives of the past in Brazil were a product of national policies, the state was also a key agent of change when it introduced affirmative action legislation, especially in the 1990s and 2000s (Htun 2004). A government turn took place under President Fernando Henrique Cardoso (1995–2003): in a combination of events that combined intellectual commitment and a post-apartheid international agenda, the 1990s in Brazil saw a growth in affirmative action in education and work opportunities (Htun 2004).

However, while the role of the state is essential to understanding both the long-lived myth of racial mixing or 'racial democracy' and more recent changes, black Brazilians still experience the state as alien (absent from the places they inhabit) or violent, and for these reasons, inclusion through state-led policies is limited. A cycle of inequality is thus easy to epitomize in Brazil. Race and the burden of slavery are at the roots of poverty and socio-spatial inequality, but a national narrative of racial democracy means that consciousness of such roots may linger unaddressed. In addition to this, whereas laws may fix categories, access to the law remains a privilege in Brazil (Holston 2009).

With the ambiguous role of the state, theorists commonly look away from policies to understand racial inclusion. Grassroots movements gain centrality to counter hegemonic narratives of racial inclusion and to promote social change. Black women, who are associated with domestic work and poorly paid jobs, and who have continuously lived in poverty, for example, will find places of leadership in *candomblé terreiros* (spaces of worship for Afro-Brazilian religion) (Perry 2016). Similarly, former gold mines can function as places of leadership for socially excluded groups in moments of spiritual manifestation. These sociopolitical spaces of memory and empowerment, however, have limits in terms of their potential to reshape experience.

Poor communities that are predominantly black and led by women struggle to access house and land tenure (Telles and Paschel 2014). Afro-Brazilian religious traditions that offer a space where black women are celebrated cannot change housing struggles. A tension between recognition in religious and cultural contexts thus exists. Food, music, sports, dance and religion feature as 'repertoires of black consciousness', but these are the same spaces that are used to promote the idea of racial democracy (Perry 2016).

Music and festivals often appear as a disputed area where either the myth of racial democracy is reborn or prejudices are unearthed. Samba can either resist cultural hegemonies or confirm social hierarchies (Chasteen 1996). The question here is whether samba, which can express racial identity, still represents a forged national identity that unifies and romanticizes Brazilians. Similar to the discussion of samba, the context of carnival, where samba is central, offers an image of society where segregation as well as counter-hegemonic narratives gain ground, contradictorily as that might sound. The choice of the parade location, the police apparatus in action and the illusion of nationality and harmony shows that the state still controls and forges the carnival's narrative, often portraying Brazil as a place of inclusion, tolerance and happiness (Baiocchi and Corrado 2010). Others disagree. For DaMatta, carnival is indeed a moment of power inversion, when a new routine is born and power dynamics are turned upside down (DaMatta 1991). According to him, carnival can be a space where society experiments with an alternative image of itself. In Ouro Preto specifically, hip-hop songs commonly speak of a city that has never changed in architectonic or social ways and about how slaves of the past are like poor black *Ouropretanos* of today.[2] Despite its powerful lyrics, the hip-hop music does not cross the threshold between peripheral areas and the city centre, and cultural music festivals in the centre do not open the space to such local groups (De Souza Santos 2019a).

Such contradictions add to my analysis in this chapter. The transformative power of mines, as discussed, has ephemeral and space-limited results. Counternarratives in mines and power dynamics of music and parades are not distinct, in that they emerge in a context of exclusion in everyday Brazilian society (Pardue 2004) but are limited in emancipatory outreach.

In mines, the inversion of power dynamics through spiritual encounters, where the spirits of slaves control the mining territory and may grant permission for visitors to enter that territory, shows a moment where spaces of violence and oppression become dominated by the oppressed. However, such an inversion is short-lived and does not move beyond the mining territory. In mines as well as at festivals, the process of counter-hegemonic narrative is incomplete. Carnival and experiences of the history of slavery are not simple racial reversal rituals. These incomplete processes of counter-hegemonic narrative are where I situate my ethnography about memories of slavery in former gold mines. I enquired why and how memories of slavery have gained traction with guided tours, and I conclude on the transformative limits of this narrative.

Conclusion

Gold mines in Ouro Preto offered an opportunity to learn about slavery through the re-creation of the working conditions in mines. The presence of slaves' spirits is explained to visitors, and spirits can affect visitors' bodies. Narratives of spirits, more than remembering, are also a form of maintaining a relationship with ancestors, as discussed in the scholarship about other colonial pasts (Cole 1998; see also: Hale 1997; Routon 2008; Shaw 2002; Stoller 1997). In this space controlled by slaves' spirits, a power inversion exists, and I problematized this power shift as short-lived and space-limited.

When Ouro Preto was nominated 'a theatre of happenings of great historical relevance' (Decree 22928, passed on 1933), the preservation of its colonial constructions followed. However, the split between the preserved centre and the improvised periphery, between the national symbolism of heroism and a local narrative that speaks about the maintenance of privileges and prejudices, and between a wealthy architecture and the neglected lives lost in the mines beneath means that there is a 'negative energy surrounding the place'. Tour guides are often amongst the residents who are spatially excluded from the cultural opportunities of the centre yet profit from the existence of a preserved city that attracts tourists. Guides commonly speak of a city that has never changed in architectonic and social ways.

A counter-hegemonic narrative in the case of Brazil is one that disputes national narratives of racial mixing, inclusive citizenship and meritocracy. Ouro Preto cements, quite literally, hegemonic national narratives (De Souza Santos and Hulme 2019). And yet, beneath the city, mines tell a story that is at once racially divided, economically unequal, ethnic and linguistically diverse, and undeniably violent.

Music, festivals and religious rituals have also been studied in contemporary Brazil as places of counter-hegemonic narratives. Instead of celebrating a racially democratic country, lyrics can denounce entrenched inequality. However, context matters. Even though the result of counter-hegemonic forms that engage with the violence of past and present may be individually or collectively positive, the narratives still emerge in a context of exclusion. For that reason, Sheriff describes carnival as a space for state hegemonic narratives (Sheriff 1999). The state controls the choice of parade location and the police apparatus; in addition, the narrative of Brazil as a plural and racially democratic nation gains popularity at such events, despite what samba may denounce and the police violence towards black participants. Even when festivals are seen as transformative (Chasteen 1996), the duration of such power reversal in society is limited.

Despite the fact that in mines colonial memories are 'stored … in flesh' (Stoller 1997), the use of mines as historic and cultural centres remains mainly connected to tourists (especially foreign tourists). Residents fear former gold mines, and tour guides, like in any other city, usually cater for external visitors.

Reckonings and the quest for justice in slavery may not go hand in hand. While remembering slavery may be possible through personal accounts with slaves' spirits, guides' dreams or visitors' experiences in mines, cultural memory remains off the record. Slavery violence, despite being no secret, is not problematized. Perpetrators of slavery in the Americas could distance themselves from the violence against slaves because of a lack of records and lack of personal slavery representations. For the victims of slavery, however, there is no such separation when economic, spatial and racial exclusion continuously pain black Brazilians. While omissions may be rectified in former gold mines, the reach of such narratives is limited by the spatial boundaries of the mines. This chapter thus supports an approach that differentiates counter-hegemonic memory from state recognition, and as such recognizes the quest for policies of racial justice. Positive achievements in terms of tours, music or festivals do exist, and they can be individually emancipatory, but more progress is necessary to address Brazil's growing racial inequality.

Dr Andreza Aruska de Souza Santos is the Director of the King's Brazil Institute, King's College London. She completed her Ph.D. in Social Anthropology at the University of St Andrews and holds a master's degree in social sciences, jointly awarded by the University of Freiburg, the University of KwaZulu Natal, in Durban, and Jawaharlal Nehru University, New Delhi. Andreza writes about Brazilian politics and city governance comparatively, and she has published widely in this area.

Notes

1. The ethnographic material below (pp. 14–21) is from my fieldwork in 2013, and it also appears in my book (De Souza Santos 2019a); I explore these passages under a new research question and literature in this chapter.
2. Chorus of song by a known hip-hop group in town:

   Rostos marcados, mãos calejadas
   Em Ouro Preto vocês podem avistar
   em cada morro da cidade o negro teve que lutar
   Pra conquistar a tua liberdade
   Que hoje no Brasil isso não é verdade
   E até hoje meu irmão não mudou nada (Rap X)
   [Marked faces, calloused hands
   In Ouro Preto you can see it
   In each hill of the city the black had to struggle to get freedom
   But freedom is not yet a reality in Brazil
   And up to now my brother nothing's changed] (Author's translation).

References

Austen, Ralph A. 2001. 'The Slave Trade as History and Memory: Confrontations of Slaving Voyage Documents and Communal Traditions', *The William and Mary Quarterly* 58(1): 229. doi:10.2307/2674425.

Baiocchi, Gianpaolo, and Lisa Corrado. 2010. 'The Politics of Habitus: Publics, Blackness, and Community Activism in Salvador, Brazil', *Qualitative Sociology* 33(3): 369–88. doi:10.1007/s11133-010-9155-z.

Batista da Costa, Everaldo. 2011. 'Totalidade Urbana e Totalidade Mundo: As Cidades Coloniais Barrocas Face a Patrimonializacao Global'. Unpublished Ph.D. thesis. São Paulo: University of São Paulo.

Cardina, Miguel, and Inês Nascimento Rodrigues. 2020. 'The Mnemonic Transition: The Rise of an Anti-anticolonial Memoryscape in Cape Verde', *Memory Studies* 14(2): 380–94. doi:10.1177/1750698020927735.

Chasteen, John Charles. 1996. 'The Prehistory of Samba: Carnival Dancing in Rio De Janeiro, 1840–1917', *Journal of Latin American Studies* 28(1): 29–47. doi:10.1017/s0022216x00012621.

Chuva, Márcia. 2009. *Os Arquitetos da Memoria: Sociogênese das Praticas de Preservação do Patrimônio Cultural no Brasil (Anos 1930–1940)*. Rio de Janeiro: Editora UFRJ.

Cole, Jennifer. 1998. 'The Work of Memory in Madagascar', *American Ethnologist* 25(4): 610–33. doi:10.1525/ae.1998.25.4.610.

Collins, J. 2011. 'Melted Gold and National Bodies: The Hermeneutics of Depth and the Value of History in Brazilian Racial Politics', *American Ethnologist* 38(4): 683–700.

Comaroff, Jean. 1985. *Body of Power, Spirit of Resistance: The Culture and History of a South African People*. Chicago: University of Chicago Press.

DaMatta, Roberto. 1991. *Carnivals, Rogues, and Heroes*. Notre Dame, IN: University of Notre Dame Press.

De Souza Santos, Andreza Aruska. 2019a. *The Politics of Memory: Urban Cultural Heritage in Brazil*, 1st edn. London: Rowman & Littlefield.

———. 2019b. 'Trading Time and Space: Grassroots Negotiations in a Brazilian Mining District', *Ethnography* 22(2): 184–206.

De Souza Santos, Andreza Aruska, and Tom Hulme. 2019. 'Cultural Marginality and Urban Placemaking', in Julie Cupples and Tom Slater (eds), *Producing and Contesting Urban Marginality: Interdisciplinary and Comparative Dialogues*. London: Rowman & Littlefield.

De Souza Santos, Andreza Aruska, and Sue A.S. Iamamoto. 2019. 'The Difficult Legacy of Mining in Past and Contemporary Potosí and Ouro Preto', *Journal of Latin American Geography* 18(3): 110–31. doi:10.1353/lag.2019.0054.

Freyre, Gilberto. 1963. *The Mansions and the Shanties: The Making of Modern Brazil*. London: Weidenfeld and Nicolson.

Fulbrook, Mary. 2020. *Reckonings*. Oxford: Oxford University Press.

Goldman, Marcio. 2013. *How Democracy Works*. Hereford: Sean Kingston.

Gonlugur, Emre, and Devrim Sezer. 2002. 'Therapeutic Forgetting, Agonistic Remembrance: Conflicting Memories of Ismir's Kulturpark and Contested Narratives in Contemporary Turkey', *Memory Studies* 14(2): 395–421. doi:10.1177/1750698020921432.

Hale, Lindsay Lauren. 1997. 'Preto Velho: Resistance, Redemption, and Engendered Representations of Slavery in a Brazilian Possession-Trance Religion', *American Ethnologist* 24(2): 392–414. doi:10.1525/ae.1997.24.2.392.

Harris, Mark. 2007. 'Ways of Knowing', in Mark Harris (ed.), *Ways of Knowing: New Approaches in the Anthropology of Experience and Learning*. Oxford: Berghahn Books, pp. 1–26.

Hatzikidi, Katerina. 2020. 'Cultural Mediators and the Protection of Ethnic Minority Cultural Rights: Reflecting on Successes and Challenges around Quilombo Heritage in Brazil', *The International Journal of Human Rights* 25(5): 752–70.

Holston, James. 2009. *Insurgent Citizenship*. Princeton, NJ: Princeton University Press.

Htun, Mala. 2004. 'From "Racial Democracy" to Affirmative Action: Changing State Policy on Race in Brazil', *Latin American Research Review* 39(1): 60–89. doi:10.1353/lar.2004.0010.

Instituto do Patrimônio Histórico e Artístico Nacional (IPHAN). n.d. 'Revista e Boletins do Instituto do Patrimônio Histórico e Artístico Nacional (1937–1947)'. Retrieved 23 October 2019 from www.docvirt.com/WI/RevIPHAN/RevIPHAN.htm.

Macdonald, Sharon. 2009. *Difficult Heritage*. London: Routledge.

Maxwell, Kenneth. 1973. *Conflicts and Conspiracies: Brazil and Portugal, 1750–1808*. Cambridge: Cambridge University Press.

Nascimento, Paulo C. 2003. 'Dilemas do Nacionalismo', *Revista Brasileira de Informação Bibliográfica em Ciências Sociais* 56: 33–53.

Owensby, Brian P. 1999. *Intimate Ironies*. Stanford, CA: Stanford University Press.

Pardue, D. 2004. 'Writing in the Margins: Brazilian Hip-Hop as an Educational Project', *Anthropology & Education Quarterly* 35(4): 411–32.

Perry, Keisha-Khan Y. 2016. 'Geographies of Power: Black Women Mobilizing Intersectionality in Brazil', *Meridians* 14(1): 94–120. doi:10.2979/meridians.14.1.08.

Rama, Angel, and John Charles Chasteen. 1996. *The Lettered City*. Durham, NC: Duke University Press.

Reed, Adam. 2002. 'City of Details: Interpreting the Personality of London', *Journal of the Royal Anthropological Institute* 8(1): 127–41. doi:10.1111/1467-9655.00102.

Ribeiro, Darcy. 2000. *The Brazilian People: The Formation and Meaning of Brazil*. Gainesville, FL: University Press of Florida.

Routon, Kenneth. 2008. 'Conjuring the Past: Slavery and the Historical Imagination in Cuba', *American Ethnologist* 35(4): 632–49. doi:10.1111/j.1548-1425.2008.00102.x.

Shaw, Rosalind. 2002. *Memories of the Slave Trade*. London: The University of Chicago Press.

Sheriff, Robin E. 1999. 'The Theft of Carnaval: National Spectacle and Racial Politics in Rio De Janeiro', *Cultural Anthropology* 14(1): 3–28. doi:10.1525/can.1999.14.1.3.

______. 2001. *Dreaming Equality*. New Brunswick, NJ: Rutgers University Press.

Souza, Jesse. 2000. 'Gilberto Freyre e a Singularidade da Cultura Brasileira', *Tempo Social* 12: 69–100.

______. 2010. *Os Batalhadores Brasileiros: Nova Classe Media ou Nova Classe Trabalhadora?* Belo Horizonte, Brazil: Editora UFMG.

Stoller, Paul. 1997. *Sensuous Scholarship*. Philadelphia, PA: University of Philadelphia.

Telles, Edward, and Tianna Paschel. 2014. 'Who Is Black, White, or Mixed Race? How Skin Color, Status, and Nation Shape Racial Classification in Latin America', *American Journal of Sociology* 120(3): 864–907. doi:10.1086/679252.

Ulturgasheva, Olga. 2017. 'Ghosts of the Gulag in the Eveny World of the Dead', *The Polar Journal* 7(1): 26–45. doi:10.1080/2154896x.2017.1329256.

– Chapter 4 –

How Industrial and Artisanal Extraction Shape a City

On Urban Planning Flaws, Encroaching Open Pits and Backyard Mining in a Congolese Mining Town

Kristien Geenen

This chapter is about a city destined to disappear. A mining city situated right on top of ore. Numerous mining companies currently exploit the mineral deposits in the city's vicinity, and their ever-expanding open pit mines are slowly but surely gnawing away at the city. In the meantime, fully aware of the peculiar location of their city on top of deposits, inhabitants secretly dig in their own backyards in search for the valuable ore. This home-bred mining, too, touches upon the foundations of the city and is making it disappear.

This city is Kolwezi, situated in the south-east of the Democratic Republic of Congo and rich in (above all) copper and cobalt. Back in the 1920s, the city's urban planning was a non-starter, as its founding fathers could not agree on where exactly the city should be situated. Indeed, there was dissension between these driving forces – the mining industry, the land registry, the politico-administrative authorities and the urbanists and geologists whom they consult – and they all seemed to have opposite goals. Simply put, while the politico-administrative bloc held a long-term view on the development of the city, the mining industry had fast and tangible results in mind. The land registry vacillated between these major contenders, looking for compromises. But as will become clear as we proceed through this chapter, the voice of the mining industry vastly outweighed – and still outweighs today – the other players, and this dominance has led to Kolwezi's uneasy location and plausible disappearance in the long run.

Figure 4.1. Aerial view of the city of Kolwezi, showing how it is hemmed in by mining activities. 2019, Google Maps.

The mining industry was initially embodied by a single company – Union Minière du Haut-Katanga, or UMHK (currently Gécamines) – and the location of the city of Kolwezi was fully determined by this company's interests. In this chapter, I show that the consequences of UMHK's insistence on building the city in a hurry, setting up makeshift buildings even before an urbanization plan was developed, are felt to this day. No thorough soil investigation was carried out; conjectures led the way. Only after the provisional buildings on haphazardly chosen plots gradually turned permanent did it become apparent that the city in the making was stuck, completely surrounded by industrial mining areas with options for future expansion. On top of this, it appeared that the city had been created directly above rich mineral deposits. Similar to the port city of Matadi (Lagae, Kivulu and Beeckmans 2016), where the peculiar topography of cliffs and slopes curb expansion, Kolwezi seemed simply to have been 'built on the wrong spot'.

Needless to say that mining seriously affects the urbanization process. Specific to mining towns is their condition of temporariness, if not of uncertainty (Ackers 1996; Calvão 2011; Emuze and Hauptfleisch 2014). Indeed, extraction develops in cycles as mining investments fluctuate with the demand on the global market. The current increased demand for cobalt, for instance, which is an important component of lithium batteries in electric vehicles, considerably raised its price and thus activities at the mines. This trend, obviously, turns downwards whenever the demand decreases. A downward trend can be reversed, however, for instance when novel metallurgical methods push mining investors to reassess the tailings that they had esteemed valueless before. In brief, different factors lead to either the (temporary) withdrawal of mining investors or to the resumption of mining activities.

These fluctuations in the mining industry equally affect the urban planning process. Following this rhythm of boom and bust, mining towns populate and depopulate; scholarly literature refers to 'shrinking cities' in the latter case (Martinez-Fernandez et al. 2012). The Congolese mining town Luena, for instance, has become virtually deserted because its coal is no longer in demand and workers left the town to look for jobs elsewhere. There are equally pop-up mining towns that spontaneously appear after artisanal miners start to dig, and these will plausibly disappear once the deposits are mined (Büscher 2018; Udelsmann Rodrigues and Tavares 2012). A complete withdrawal of mining activities leads to 'zombie mines' that ravage the natural landscape and need to be reimagined or given new functions (Goin and Raymond 2004; Iannella 2014; Keeling and Sandlos 2017). All these cases illustrate the (eventual) temporariness and vicissitudes that accompany the extractive industry.

Mining impairs the urban outlook too. It is above all the extraction of copper through open pit mines that leads to the typical minescapes, with ever expanding mines that leave a gaping hole in the face of the earth (Francaviglia 1991; Nash 1979). As such, mining inevitably triggers an interaction with the inhabitants. In his eloquently named book *The City That Ate Itself*, Brian James Leech (2018) outlines how the creeping borders of the Berkeley Pit in the city of Butte in Montana not only affected the urban but also the social tissue of the city. As the mining pit grew larger, forced removals of whole neighbourhoods were rife. In a similar vein, Janet Finn (1998) demonstrated how not only Butte but also the mining city Chuquicamata in Chile have been put to the test by the expansion of mining activities. The authors explored how the communities reacted to the encroachment of their city and highlighted the inhabitants' resilience. The latter are indeed forced to react because the physical reminders of the former mining days simply cannot be ignored but are omnipresent in the cityscape and often even crucial to its urban planning. The Congolese city of Manono, which was entirely dependent on tin mining, clearly illustrates how the mining industry is decisive for the urban outlook even long after its withdrawal. Indeed, the dam that was constructed to suit the mining industry still splits the city in two, and its former mining camps with monotonously rectangular houses mark the cityscape to this day (Bruneau and Mukalayi Kahata 1982). To the post-mining communities, all these physical remains evoke the remembrances of past mining times while they search for alternative means of subsistence in the present and keep up hope for a mining revival in the future. The totality of these mining legacies and the resource extraction that produces the global minescape are under scrutiny in this book (see Introduction).

The city that this chapter will discuss, Kolwezi, is such a product of the mining industry that it conjures up 'a science-fiction landscape' (Bruneau and Mansila Fu-Kiau 1986). It is a place where mining continuously pushes inhabitants to (be prepared to) move as the open mine pits encroach upon the city; a place that is likely to be removed and relocated altogether. As I will argue, it was – and still is – the interests of the mining industry that guide the urbanization process of Kolwezi, contravening in the process the ground rules of urban planning. As such, mining is the source not only of Kolwezi's creation but also of its plausible destruction. The main characteristic of the city seems to be its temporality. In a paradoxical way, the only constancy of Kolwezi is its imminent disappearance; a spatiality and temporariness that Achille Mbembe (2014) refers to as 'the zero world'. It is at once a gem of the current mining boom, which exudes a bright future, and a place that appears doomed.

And yet, the inhabitants of the city – called *Kolwéziens* – make as much as they can of the situation by digging up mineral ore in their backyards and selling it to the local trading houses. At the end of the global chain, this cobalt ends up in devices such as mobile phones and electric bikes. Although conditions are harsh and unsafe, these home-bred activities do allow some city dwellers to make ends meet. In doing so, they adjust to their peculiar urban and geological context and use it to sustain their personal advancement. By self-generating the livelihoods they expect their mineral-rich soil to bring forth, Kolwezi's inhabitants respond with genuine inventiveness to the peculiar urban features of their city. However, as I will argue in this chapter, by digging ore in their backyards, they are at the same time undermining the very foundations of their homes and making nearby roads collapse. Together with the activities of the mining industry, they are making their city disappear.

The chapter is divided into two main parts, one situated in the past, and the other in the present day. I first outline the historical context of the city's development and illustrate that the logics of capital accumulation were decisive for the way it grew. In the second part, I explore how the current inhabitants of Kolwezi deal with the result of this entanglement with the mining industry and urban planning, in living on top of mineral deposits. My argument emerged from a close reading of the correspondence between representatives of UMHK, the Belgian colonial authorities at the African Archives in Brussels and from recent interviews with city dwellers, experts and authorities in Kolwezi.[1]

Laying Cornerstones with a Sense of Urgency

That the city of Kolwezi was built far too hastily was due to its belated connection to a railway network. Already at the turn of the nineteenth century, different teams of experts, under the aegis of both Leopold II's Congo Free State and Robert Williams's Tanganyika Concessions Limited (TLC), had pinpointed the deposits of Ruwe, Kolwezi and Musonoi, all three situated in the largely uninhabited savannah where the city of Kolwezi would be established.[2] The very first labour camp was set up near the Ruwe deposit in October 1903, accommodating five Europeans and some hundred workers,[3] and the other two deposits too triggered human settlement. The beginnings of production then followed. But the site was rather isolated and options for export were few. In 1904, Robert Williams ordered the opening up of a route for wagons between Ruwe and Benguela at the Atlantic coast in Angola. An experiment in 1906 proved that it took an oxcart as long as eight months to transport the mineral ore and ingots from this area to the Angolan port for shipping, hardly a viable production line.[4] Meanwhile, also in 1906, TLC and Comité Spécial du Katanga (CSK), a semi-private concessionary company that administered the land and was created in 1900, launched the mining company UMHK. Unsurprisingly, when in 1910 UMHK's other two centres of activity,

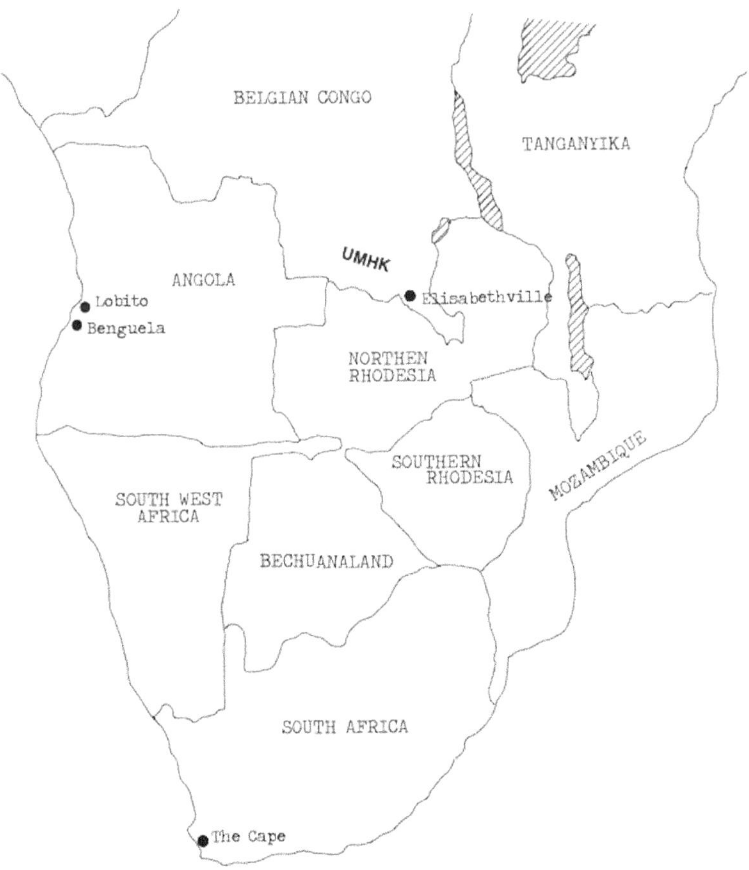

Figure 4.2. Map of Africa, 1930s, Brigit Willems.

Elisabethville with the Etoile du Congo mine and Panda with the Kambove mine, were connected by railway to Sakania and destinations further south towards the Cape, UMHK shifted its focus to this so-called Groupe Sud and Groupe Centre, at the expense of Groupe Ouest with the Ruwe-Kolwezi-Musonoi deposits (on the railway network, see Katzenellenbogen 1973; Hance and Van Dongen 1956).

At the beginning of the 1920s, when a railway connection between the port of Benguela and the future town of Kolwezi was finally in the making, UMHK resumed prospecting at Groupe Ouest, particularly at the deposits of Ruwe and Musonoi.[5] These mining activities attracted a growing population, and the idea then arose that a town should be established nearby to accommodate all these immigrants (Chapelier 1956a: 299). Hesitantly, urbanization took off. The first built areas consisted of the labour camps that UMHK constructed near the mines to accommodate its workforce. As Bruno de Meulder (1996) has shown, these camps combined temporariness with rationality. In his words, 'it is precisely because of its temporary nature that the camp reveals a basic rationality: it is an unequivocal solution, free of additional duties, to a given problem' (De Meulder 1996: 8).[6] The given problem was to ensure the availability of a sufficient number of competent and healthy workers, and accommodating the workers next to their place of work seemed a foolproof solution.

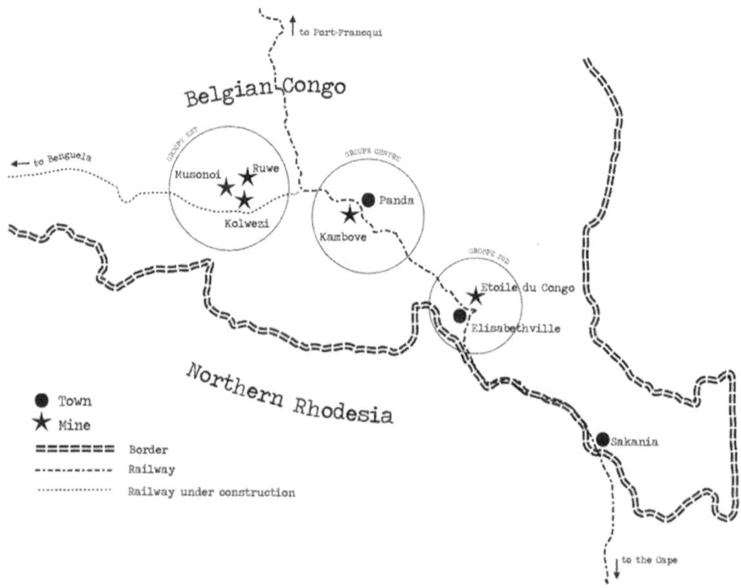

Figure 4.3. The three Groups of UMHK in the Congolese Copperbelt, 1930s, Brigit Willems.

But UMHK failed to take future expansions of both the mines and the camps into account, with unfortunate results. On top of this, the design of the camps was in the hands of UMHK's physicians, Mottoullc and Van Nitsen, who, obviously, placed a greater emphasis on health considerations than on urban planning standards (De Meulder 1996: 119).

UMHK and CSK closely cooperated. CSK administered the land and owned the deposits identified at that stage, which it ceded to UMHK. It also intervened if land tenure conflicts arose between UMHK and local traditional chiefs. UMHK had the exclusive surface rights for prospecting and pegging those deposits as yet unidentified on the immense land it had been accorded. On top of this, CSK had allotted the company a considerable amount of land to set up its future mining and metallurgical installations, all situated within the mining polygon. Besides UMHK, the Belgian colonial administration, the railway company Chemin de Fer du Bas-Congo au Katanga (BCK) and CSK itself were entitled to free plots on which to build their offices and houses for their personnel. To accommodate the few European colonists and traders who were drawn to the budding town, CSK set up a separate allotment of plots in 1921 (Chapelier 1956b: 63; Henk 1988: 154). Called camp Dilala, this area, however, was supposed to remain temporary, as it was situated 'in an area that will be annexed to the mining polygon' (Bruneau and Mansila Fu-Kiau 1986: 220); tenancy agreements needed to be renewed each year and CSK had the right to recover the land at any time (Chapelier 1956a: 299). Obviously, the labour camps accommodated a far more numerous population and thus shaped the urban landscape far more decisively from the outset. As we have seen, UMHK built these labour camps as close as possible to mines and in non-durable materials. A close look at the correspondence between the driving forces identified at the beginning of this chapter clearly shows that they proceeded initially in a somewhat haphazard fashion, seemingly averse to basic urban planning principles. The provisional prevailed; permanent solutions were postponed. For one thing, there was no master plan; proximity to a centre of activity was the sole criterion for settlement. Since in the beginning mining activities demanded a considerable operational workforce, UMHK emphasized the urgent need to prioritize construction despite CSK's reservations. At this point, all the bodies initially involved made an error of judgement in grossly underestimating the future growth of the city. Indeed, a 1921 report stated that 'one should not exaggerate the apparent importance of Musonoi as most of the plots will never be occupied' (as cited in Chapelier 1956a: 61–62). Nevertheless, tensions between the driving forces soon showed. When CSK suggested that a specialized organization should draw up a plan for the city *first*, and *then* incorporate additional Europeans,[7] UMHK responded with a request that the choice of the location of the future city be accelerated and implemented immediately.[8]

The hastily constructed labour camps destined to be temporary turned out to be permanent. Indeed, pushed by personnel shortages, UMHK tried a new tack and in 1926 launched a campaign to stabilize its workforce 'in order to push the black worker to love his work and stick to it as long as possible' (Mottoulle 1946: 11). To make mining work more attractive, the management traded the usual short contracts for longer ones lasting three years and renewable thereafter. Part of the deal consisted of better dwellings, too, built with more sophisticated and durable materials, enabling the worker to settle with his wife and children (on the stabilization policy, see Banjikila Bakajika 1993; De Meulder 1996; Dibwe dia Mwembu 2001; Fetter 1973; Perrings 1979; Rubbers 2013). The outlook of the labour camps changed considerably, as schools and playgrounds were added, as well as hospitals, sports facilities and even lines of trees to offer shade. Workers were allowed to add a personal touch to their homes, by painting the walls or growing flower gardens, which was previously forbidden. The coercive look of the first labour camps softened into the more convivial look of workers' neighbourhoods (*cités ouvrières*). With a steadily growing population, the need for urbanizing the area became more urgent. The future town was to arise somewhere 'within our exploitation and mining polygon', as the head of UMHK put it.[9] Note that all these issues were discussed behind closed doors *before* CSK and UMHK informed the colonial governmental bodies, either in Congo or in the Belgian metropole. When finally this third party was brought on board and a reunion was settled with provincial Governor Heenen in June 1929 to consider the location of the future city, all that UMHK did was reinforce the state of urgency and state its wish to start construction without further ado.[10]

When the very same year a first draft of the future city was tabled, René Detroy, a Belgian urbanist, wrote a scathing criticism of the rush that had engulfed the project. In his words, 'the draft of the project is not elaborated enough what concerns the allotments, the roadworks, the buildings, the aesthetics and the order of urgency'. It is particularly this last remark, the lack of a sound list of priorities, that reveals the tensions between the city's founding fathers. UMHK clearly wanted to get started as soon as possible, unhindered by any long-term vision of urban development. CSK, on the other hand, wanted to plan the 'basic layout of the city' first, before the erection of any building. René Detroy supported this idea, which was clearly more in line with the ground rules of urbanization. UMHK's ideas had collided with those of CSK and of the colonial authorities, advised by a Belgian urbanist. The Belgian authorities in the metropolis, for their part, complained vociferously that their authority had been bypassed. In February 1930, the Ministry of Colonies in Brussels warned that it was *dangereux* ('dangerous') to proceed without its approval, further denouncing the refusal of UMHK, despite several requests, to present a detailed map of the future Kolwezi.[11] Nevertheless, UMHK won the day. Barely six months later the Belgian authorities approved the final convention between UMHK and CSK concerning the creation of the urban agglomeration.[12] Thereafter, urban development hit the ground running when in 1931 the Benguela Railway finally linked the region to the port of Benguela on the Atlantic, and to the port of Lobito 35 km further north. Evidence suggests that Kolwezi then entered into a kind of competition with UMHK's other two centres of activity, Elisabethville and Panda-Likasi; in order to catch up, UMHK rushed its fences and – consequently – the area was urbanized with great speed. UMHK (and to a lesser extent other companies such as BCK) was attracting a growing number of European employees, along with additional administrators and retailers to assure their civic rights and service their needs. Camp Dilala had by then been dismantled,[13] but CSK offered the expanding retail sector the opportunity to buy plots in a newly created small commercial centre; by 1936, six traders had done so.[14] The idea then arose that it was time to streamline the budding city, since – regardless of the existing urban plan – it was clear that it still lacked coherence. Akin Mabogunje (1990: 139) has shown that it is not an unusual feature for a colonial city to be spread out, but rather the result of 'a fragmented civil society [which] ensures a selective approach to urban planning'. The European administrators, the traders, the local workforce, the immigrants, they were all accommodated in separate neighbourhoods, with a neutral zone keeping the Africans out of the lives of the Europeans, and vice versa. In future Kolwezi, however, it was the distance between the three deposits, Ruwe, Kolwezi and Musonoi, that led to scattered centres of activity. As we have seen, UMHK had regrouped its labour camps and buildings around these three sites, and consequently the town remained amorphous. 'What strikes the visitor to this area is the eccentric location of its different centres of activity', was the verdict of Putmans, who added that without any doubt this was the result of the over-hasty opening up of the mines of Groupe Ouest for exploitation.[15] The Belgian authorities had ordered the mining engineer to survey the site in 1936. Time had come to choose the central location of the city, a site to tie the urban fabric together, and expand it outwards from there. As the carrière de Kolwezi was situated roughly halfway between Ruwe and Musonoi, in December 1938 UMHK chose the nearby Plateau de la Manika to house the first buildings. This site was very convenient, as there were already some good quality houses on this plateau (at the Square des Araucarias), which were destined for unmarried European workers of UMHK and a smaller number of merchants working in the nearby commercial centre; CSK and Cofoka therefore agreed to extend the future city from this square outwards (Chapelier 1956a: 302; Henk 1988: 156). Accordingly, in 1939, CSK started to allocate plots in this newly founded centre to private individuals too.[16]

The year 1937 was the start of an important production phase as mechanized extraction took off (Perrings 1979: 122) and the concentrator was installed (Henk 1988: 156). As word spread that Kolwezi was the new hive of opportunity, jobseekers came in droves, often settling in the nearby village of Mwangeji. To channel this stream of so-called 'uprooted' ones (*'les déracinés'*), the authorities planned an indigenous neighbourhood (*cité indigene*), which was also meant to house the additional workforce of BCK and Cofoka that could find no room in the existing camps.[17]

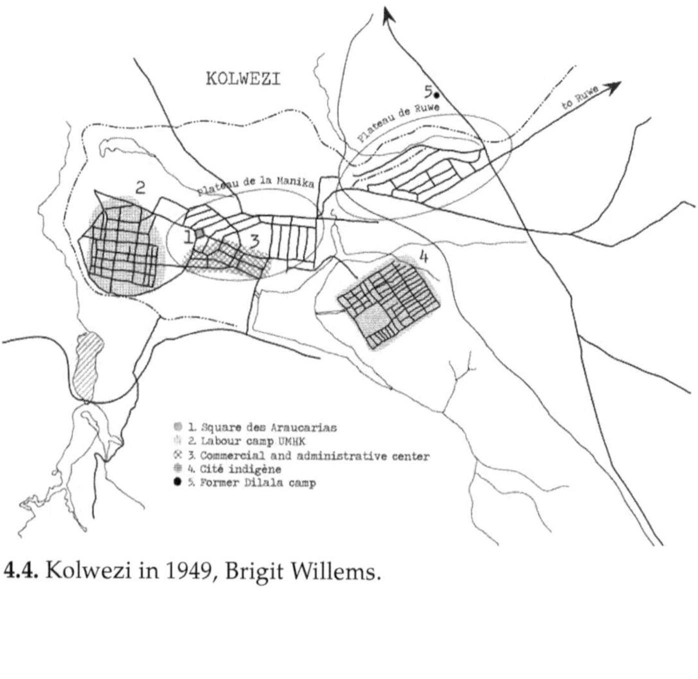

Figure 4.4. Kolwezi in 1949, Brigit Willems.

The consequences of the precipitous choice of location for the built areas – no decent plan, nor thorough soil examinations – soon surfaced. As it turned out, buildings were erected on land that UMHK would have preferred to exploit. Already in 1940, nearly a year after the birth of the city, the homes of two territorial agents in the Musonoi neighbourhood needed to be destroyed, since, as the company stated, they were situated 'right on top of a mineral deposit'.[18] The town's location turned problematic as the mining industry gradually engulfed the town; it seemed to be hemmed in by open mines, plants and a concentrator but lacked an urban plan to straighten this out. In 1947, the principal territorial agent, Henroteaux, expressed his concern and stated that urgent measures were warranted.[19] This clarion call did not fall on deaf ears. The very same year, the Belgian authorities sent architect-urbanist Noël Van Malleghem to the Belgian Congo to survey the area. By that time, as we have seen, Kolwezi consisted of different kinds of urban areas, as was common in other Congolese cities: the small quarter for European inhabitants, which included the commercial and administrative area; the different labour camps situated in the vicinity of the mines; and lastly the *cité indigène*, an area reserved for fresh immigrants, separated from the European quarter by a neutral zone. UMHK's labour camps still largely dominated the area. A new decree on urbanism in 1949 demanded that every time a neighbourhood in a Congolese town was expanded or modified a thorough plan should be drawn up and approved by the administrative authorities (Chapelier 1956a: 305). Every future expansion of Kolwezi was going to be strictly controlled, and this was precisely Van Malleghem's assignment. He was particularly harsh on the messy ensemble that Kolwezi still presented, 'by lack of a holistic plan'. He outlined two major objections, the omnipresence of the mining industry and the unfortunate location of the city; as he put it in his report, 'we would like to stress how inconvenient it would be if Kolwezi were allowed to expand at its current location'. In his view, the hurried proceedings had led to band-aid solutions[20] and to a place where 'houses are literally stuck to the industry'.[21]

Time had come to turn over a new leaf. It is at this stage that for the very first time the idea arose to simply remove the city and relocate it elsewhere. Van Malleghem suggested the founding of a new town on the nearby Plateau de Ruwe, a town he referred to as 'the second Kolwezi' (see Figure 4.4). [22] Just a year later, CSK indeed created 260 new plots on this plateau, not to find a new place disconnected from the mining industry as Van Malleghem advocated but

Figure 4.5. Aerial view of UMHK's labour camp and the mines nearby, 1949. Algemeen Rijksarchief, Brussels, 3DG(200)1: 'Rapport de mission' by Van Malleghem, 1949.

simply because the Plateau de la Manika was 'full'. As such, after the establishment of this 'second Kolwezi' (Chapelier 1956a: 306), the former town at Plateau de la Manika remained right where it was, malpositioned. And as we will discuss shortly, the newly chosen Plateau de Ruwe turned out to be rich in minerals too. Indeed, it seems that the entire city is built upon deposits.

Local Entrepreneurship

With previously unexamined land and consequently undiscovered deposits all over, a repositioning of the whole city seems the only viable solution in the long run. And this removal appears to have been pending for a long time. Gaston, a centenarian and inhabitant of the very first labour camp near the Kolwezi mine, now called cité Gécamines,[23] remembers that the Belgian head of the camp regularly passed by the inhabitants to warn them that they should move as their houses would be destroyed to make room for mining activities.[24] Marie-Paule Dutordoir, who worked as a social assistant at UMHK, recalls that from the time she arrived in Kolwezi in 1957 onwards, there were talks of the city's relocation.[25] As mining rights always override land rights, the *Kolwéziens* got used to unexpected turns and to living with the threat of removal, feeling as if they were on borrowed time. The only constancy of the city was the idea that it would disappear.

The authorities undertook the first serious attempt to move the entire city in 1972. According to this plan, Kolwezi was to be relocated 20 km eastwards to the fringes of lake Nzilo, away from the industrial sphere to make it more *city-like*. However, the removal never eventuated, blocked by lack of funding (Bruneau and Mansila Fu-Kiau 1986).[26] But whereas a removal of the entire city is perhaps unrealizable, smaller parts have gradually disappeared. Quite recently, the Sino-Congolese joint venture Compagnie Minière de Musonoï (COMMUS) picked up the thread of activity in the Kolwezi mine, which had lain dormant since 1963.[27] Gradually, all the homes on the fringes of *cité* Gécamines were affected, and the inhabitants were pressured to move after all, without compensation. Nevertheless, and similar to what happened during colonial times, mining companies take decisions in a spirit of *we shall see when the time comes*. A provincial government official said that when managers were asked what solution they had in mind for the threatened inhabitants, they simply replied that they would find a solution when they started mining.[28]

Procrastination by the mining industry seems to have been a leitmotif throughout the urbanization of Kolwezi, triggered by an everlasting hurry. As we have seen, inhabitants live with the idea that they might be expelled from their homes to make room for mining at any time. But where to go? It seems that the whole city is built upon rich mineral deposits; without irony, *Kolwéziens* speak of 'the hospital's deposit' (*'le gisement de l'hopital'*). Sooner or later, all these built-up areas that block access to mineral deposits will have to disappear. But where will the *Kolwéziens* settle?

Let us now consider the present time. How to lead a life living on deposits that seem to be everywhere? What happened in Kasulo is very illustrative in this respect. Kasulo is one of several neighbourhoods with a more organic and uncontrolled growth that was infringing on one of Gécamines' mineral-rich polygons (Mansila Fu-Kiau 1988: 277). By the mid-1980s, the existing labour camps were dwarfed by these newer neighbourhoods, which occupied nearly 50 per cent of the residential area and accommodated about 30 per cent of Kolwezi's inhabitants (Bruneau and Mansila Fu-Kiau 1986: 225; Mansila Fu-Kiau 1988: 65). A couple of years ago, one resident dug a hole in his tiny compound in order to create a septic tank. While digging, he found that his penurious plot had unexpected bonuses to offer: it appeared rich in cobalt ore. He started selling it to traders, often Chinese, at the trading houses (*comptoirs*) at nearby Musompo. He soon made a small fortune, enabling him to move out and settle comfortably elsewhere. Neighbours were inspired by his stroke of luck and started digging in their compounds too. Before long, the whole Kasulo neighbourhood turned into a buzzy mining area. With their mallets and chisels, inhabitants dug as deep as forty meters to access the ore, tunnelling underneath the whole neighbourhood. To protect them from the rain, they covered the pits with orange canvasses. Some even dug *inside* their houses (easy to do, as the floors consist of nothing but flattened earth) in order to better shield the pit and keep the curious away. Not everyone had the right skills to dig though, and many inhabitants allowed more experienced miners to dig in their compounds. Usually, a team consisted of five *creuseurs*, exclusively male, mining being a prominently masculine activity in the Congolese Copperbelt (Pugliese 2020). Among them, 40 per cent of the revenues went to the inhabitant who gave access to the compound, 40 per cent to the five *creuseurs* and 20 percent to their foreman.[29] The first couple of years, the municipal and provincial authorities tried to curb these artisanal activities, but failing to provide the *creuseurs* with an alternative site, they chose to turn a blind eye.[30] By the end of 2017, however, pushed

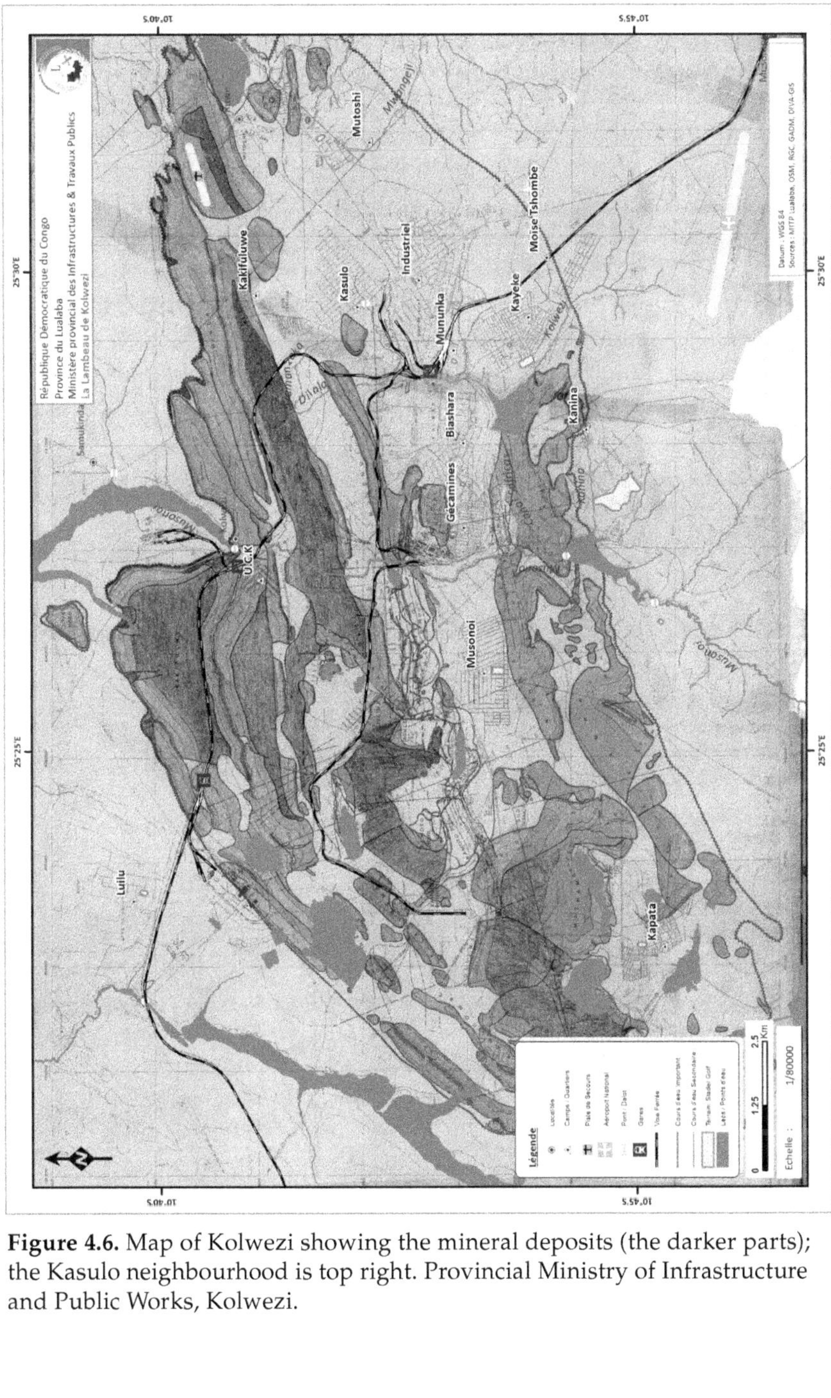

Figure 4.6. Map of Kolwezi showing the mineral deposits (the darker parts); the Kasulo neighbourhood is top right. Provincial Ministry of Infrastructure and Public Works, Kolwezi.

by the findings of an Amnesty International report (2016) and fearing a negative image, they proclaimed the manual digging illegal, falling back on the argument that the neighbourhood was a residential area after all. Their main incentive, so they stated, was the poor labour conditions, endangering the lives of numerous children active on the site (Kayumba 2017). Gécamines, which still holds the Permit of Exploitation (PE) of this neighbourhood, granted the concession to the Chinese company Congo Dongfang International Mining (CDM), and six hundred families were asked to move out, without compensation.[31] CDM put a high wall around the neighbourhood, and the Ministry of Mines turned its statute into a Zone for Artisanal Mining. CDM today engages some of the former inhabitants as well as about 7,000 *creuseurs* to continue to dig into, as one framed it, 'his pit', but the miners need to sell all minerals to the company exclusively.[32]

Figure 4.7. Two aerial views of the Kasulo neighbourhood. The top image from 2017 shows the orange canvasses that the inhabitants used to cover up the pits in their backyards. The bottom image from 2021 shows the wall built by CDM, and the buildings of the company at the northern side within the walled compound. Google Maps.

Figure 4.8. CDM's wall around the Kasulo neighbourhood, 2019. Photo by Felly Kusaya Mushisa.

The story ends with a fascinating turn. By means of compensation, the authorities allotted a new neighbourhood to the former inhabitants of Kasulo. Interestingly, the decision makers once again made the very same mistake, acting in a hurry, with the mining industry again in the vanguard. The committee that decided where the inhabitants would be relocated did not include an expert on urbanism. Rather, the Ministry of Mining held a prominent place alongside employees of the cadastre. One member of the committee was flabbergasted by the way the new area was chosen, without any preliminary exploration of the soil, and without any regard for urbanist criteria.[33] Instead, the employees of the cadastre pointed out a site that was vacant, and the other committee members agreed that 'this site looks nice'. So when the inhabitants of Kasulo settled in their newly created neighbourhood, Joli Parc in Samukinda, their fingers were itching to start digging anew. They soon encountered yet another windfall: their new compounds too appeared to be rich in cobalt ore.

By digging underneath their own houses, *Kolwéziens* are simultaneously generating an income and undermining the very foundations of their homes. Similar to the activities of the mining companies, their backyard-digging is making the city disappear. I have shown that the interface between the mining industry and urban planning has been uneasy. However, the rhythm of Kolwezi's growth is self-destructive – malpositioned as it is – and its urbanization is self-defeating. As Alice Chapelier (1956b: 50) observed in 1956, the harmonious development of Kolwezi was 'compromised' by the occupation of a site that had been slavishly chosen in function of the mining industry. In Kolwezi, in a very gradual way, the built space has been revealed to have encroached on the mineral deposits, with the city having been 'built on the wrong spot'. The city is gradually being eaten away, not only by the mining industry but also by the inhabitants, as the impromptu artisanal mining activities in the Kasulo neighbourhood have shown.

Kristien Geenen holds a Ph.D. in Social and Cultural Anthropology (University of Leuven, promotor Filip De Boeck). Her fieldwork is long-standing in urban DR Congo, specifically in the cities of Kinshasa, Butembo and Kolwezi, and she has a keen interest in historical research. She was affiliated as a postdoctoral researcher to Ghent University (interdisciplinary project 'Urban Landscapes of Colonial/Postcolonial Health Care', supervisor Johan Lagae) and to Université de Liège (research project 'Work in Mining', supervisor Benjamin Rubbers). She is currently affiliated to the research project 'Résolution Métis', supervisors Pierre Alain Tallier and Nico Wouters.

Notes

1. General acknowledgements: I warmly thank Filippo M. Zerilli and Antonio Maria Pusceddu for organizing and inviting me to the Global Life of Mines event held in Cagliari, November 2019, and for offering me the opportunity to present the current chapter. This chapter was written as part of the WORKinMINING project (www.workinmining.ulg.ac.be). The project has received funding from the European Research Council (ERC) under the European Union's Horizon 2020 research and innovation programme (grant agreement no. 646802). The ideas developed in this chapter reflect only the author's view. The ERC is not responsible for any use that may be made of the information it contains.

   A brief discussion of toponyms is warranted here. Before Kolwezi was called into being, there were three deposits nearby, called Ruwe, Musonoi and Kolwezi. The name Kolwezi was not used to refer to the urban area until December 1939, when the colonial authorities accorded the place the statute of urban district (*circonscription urbaine*); before that, the name Kolwezi referred to the river and the deposit exclusively. To refer to the place, different names were in use at the same time, often related to one of the other two deposits, Ruwe or Musonoi. At some stage, Jadotville was in the running too, but in 1930 this name was accorded to the city created close to UMHK's production centre Panda (currently Likasi) (Chapelier 1956a: 300; Henk 1988: 155).
2. Note that 'practically all [deposits] were on sites of old mines' where the local population was active and that the mine of Kolwezi was still operating by 1899 (Perrings 1979: 7–8). As such, '[p]rospecting in Katanga was reduced to locating "ancient workings" rather than using scientific exploration methods to discover previously unknown ore bodies' (Särkkä 2016: 330). Indeed, the Sanga ethnic group were mining and smelting copper long before the European expeditions (see Cuvelier 2011; De Hemptinne 1926; Gobbers 2016; Mukembe Mubedi 2018).
3. African Archives, Brussels (henceforth AA) AGR2/1269: 'Gisement Auro-Platinifère de Ruwe. Résumé de la documentation existente par E M Richet'.
4. AA AGR2/1269: 'Gisement Auro-Platinifère de Ruwe. Résumé de la documentation existente par E M Richet'.
5. AA AGR2/1269: 'Gisement Auro-Platinifère de Ruwe. Résumé de la documentation existente par E M Richet'.
6. All translations by the author.
7. AA 3DG(824)3: letter from head of CSK Gohr to head of UMHK, 14 May 1929.
8. AA 3DG(824)3: letter from head of UMHK to head of CSK, 12 June 1929.
9. AA 3DG(824)3: letter from head of UMHK to head of CSK, 12 June 1929.
10. AA 3DG(824)3: report of the meeting on 26 June 1929.
11. AA 3DG(824)3: letter by Ministère des Colonies to secrétaire générale of CSK, 12 February 1930.
12. AA T(114), 371: 'Convention entre le Comité Spécial du Katanga et l'Union Minière du Haut-Katanga pour l'établissement d'une agglomeration urbaine dans la region des mines de l'Ouest', 11 August 1930.

13. Camp Dilala was restored in 1949. Private traders such as Bozzone and Forrest housed their personnel at this camp (AA GG 17103, report 'Contrôle du poste administratif de Kolwezi, 1949' by Roland, 1949).
14. AA 3DG(1750)10: 'Rapport d'inspection de la mine de cuivre de l'UMHK à Musonoie' by E. Putmans, 6 July 1936.
15. AA 3DG(1750)10: 'Rapport d'inspection de la mine de cuivre de l'UMHK à Musonoie' by E. Putmans, 6 July, 1936.
16. AA RF(1716): report of meeting CSK on 7 July 1939, titled 'Création d'un lotissement public à Kolwezi'.
17. AA Rapport AIMO 1946 by L. Henroteaux, administrateur territorial principal, 29 January 1947. By 1946 the *cité indigène* was finished, and about twelve hundred inhabitants had settled there.
18. AA 3DG(138)4: report of meeting in Brussels between colonial authorities, CSK, UMHK and Cofoka, 17 April 1940.
19. AA Rapport AIMO Kolwezi 1946, L. Henroteaux, administrateur territorial principal, 29 January 1947.
20. AA 3DG(186)1: 'Avant projet du plan général d'aménagement' by Van Malleghem, 1947.
21. AA 3DG(200)1: 'Rapport de mission' by Van Malleghem, 1949.
22. AA 3DG(189)7: letter by Van Malleghem to Ministry of Colonies, 29 August 1951.
23. All former labour camps are presently called *cité*; *cité* Gécamines is situated close to the Kolwezi mine, *cité* Musonoi and *cité* Mutoshi close to the Musonoi and Ruwe mine respectively.
24. Gaston, interview with the author, Kolwezi, 22 November 2017.
25. Marie-Paule Dutordoir, interview with the author, Kolwezi, 30 March 2017.
26. The International Bank for Reconstruction and Development was supposed to fund the removal (Mansila Fu-Kiau 1988: 79).
27. COMMUS is a 73/27 joint-venture between China Overseas Engineering Group Corporation Limited (COVEC) and Gécamines.
28. Provincial authority, interview with the author, Kolwezi, 2 December 2017.
29. Anonymous *creuseur* 1, interview with the author, Kolwezi, 30 June 2018.
30. See 'Kolwezi: Artisanal Miners Ordered to Stop Mining in Kasulo'. Available at: www.radiookapi.net/actualite/2014/07/24/kolwezi-les-creuseurs-artisanaux-sommes-darreter-lexploitation-miniere-kasulo.
31. CDM is a 100 per cent owned subsidiary of China-based Zhejiang Huayou Cobalt Company Ltd, one of the world's largest manufacturers of cobalt products.
32. Tati, interview with the author, Kolwezi, 28 January 2018.
33. Geologist, interview with the author, Kolwezi, 22 February 2018.

References

Ackers, Peter. 1996. 'Life after Death: Mining History without a Coal Industry', *Historical Studies in Industrial Relations* 1: 159–70.

Amnesty International. 2016. *This Is What We Die For: Human Rights Abuses in the Democratic Republic of the Congo Power the Global Trade in Cobalt*. London: Amnesty International.

Banjikila Bakajika, Thomas. 1993. 'Capitalisme, Rapport Salarial et Regulation de la Main-d'oeuvre: La Classe Ouvrière Noire dans les Camps de l'Union Minière du Haut-Katanga, 1925–1967', Ph.D. dissertation. Université de Laval.

Bruneau, Jean-Claude, and Mukalayi Kahata. 1982. 'Les Paysages Urbains de Manono (Zaïre)', *Cahiers d'outre-mer* 140: 363–74.

Bruneau, Jean-Claude, and Shourit Mansila Fu-Kiau. 1986. 'Kolwezi: L'Espace Habité et ses Problèmes dans le Premier Centre Minier du Zaïre', *Cahier des Sciences Humaines* 22(2): 217–29.

Bryceson, Deborah, and Danny MacKinnon. 2012. 'Eureka and Beyond: Mining's Impact on African Urbanisation', *Journal of Contemporary African Studies* 30(4): 513–37.

Büscher, Karen. 2018. 'African Cities and Violent Conflict: The Urban Dimension of Conflict and Post Conflict dynamics in Central and Eastern Africa', *Journal of Eastern African Studies* 12(2): 193–210.

Calvão, Filipe. 2011. 'When Boom Goes Bust: Ruins, Crisis and Security in Megaengineering Diamond Mining in Angola', in Stanley D. Brunn (ed.), *Engineering Earth: The Impacts of Megaengineering Projects*. New York: Springer, pp. 367–82.

Chapelier, Alice. 1956a. 'Elisabethville, Jadotville et Kolwezi: Etude de Géographie Urbaine Comparée', Ph.D. dissertation. Université de Liège.

———. 1956b. 'Elisabethville, Jadotville et Kolwezi: Etude de Géographie Urbaine Comparée. Annexes', part two of Ph.D. dissertation. Université de Liège.

Cuvelier, Jeroen, 2011. 'Men, Mines and Masculinities: The Lives and Practices of Artisanal Miners in Lwambo (Katanga province, DR Congo)', Ph.D. dissertation. Ghent University.

De Hemptinne, Jean-Félix. 1926. 'Les Mangeurs de Cuivre au Katanga', *Congo* 1(3): 371–403.

De Meulder, Bruno. 1996. *De Kampen van Kongo: Arbeid, Kapitaal en Rasveredeling in de Koloniale Planning*. Amsterdam: Meulenhoff.

Dibwe dia Mwembu, Dontatien. 2001. *Bana Shaba Abandonnés par leur Père: Structures de l'Authorité et Histoire Sociale de la Famille Ouvrière au Katanga 1910–1997*. Paris: L'Harmattan.

Emuze, Fidelis, and Cornel Hauptfleisch. 2014. 'The Impact of Mining Induced Urbanization: A Case Study of Kathu in South Africa', *Journal of Construction Project Management and Innovation* 4(2): 882–94.

Fetter, Bruce. 1973. *L'Union Minière du Haut-Katanga, 1920–1940: Naissance d'une Sous-Culture Totalitaire*. Les Cahiers du CEDAF 1976/6.

Finn, Janet. 1998. *Tracing the Veins: Of Copper, Culture, and Community from Butte to Chuquicamata*. Berkeley, CA: University of California Press.

Francaviglia, Richard. 1991. *Hard Places: Reading the Landscape of America's Historic Mining Districts*. Iowa City, IA: University of Iowa Press.

Gobbers, Erik. 2016. 'Ethnic Associations in Katanga Province, the Democratic Republic of Congo: Multi-tier System, Shifting Identities and the Relativity of Autochthony', *Journal of Modern African Studies* 54(2): 211–36.

Goin, Peter, and Elizabeth C. Raymond. 2004. *Changing Mines in America*. Santa Fe: Center for American Places.

Hance, William A., and Irene S. van Dongen. 1956. 'The Port of Lobito and the Benguela Railway', *Geographical Review* 46(4): 460–87.

Henk, Daniel W. 1988. 'Kazi ya Shaba: Choice, Continuity, and Social Change in an Industrial Community of Southern Zaïre', Ph.D. dissertation. University of Florida.

Iannella, David. 2014. 'Arizona's "Zombie" Uranium Mines: Lax Regulations Threaten Los Angeles Tap Water', *Boston College Environmental Affairs Law Review* 41(3): 54–67.

Katzenellenbogen, Simon E. 1973. *Railways and the Copper Mines of Katanga*. Oxford: Clarendon Press.

Kayumba, Olivier. 2017. 'Kasulo et Tshipuki Fermeture des Carriers pour le Salut des Enfants', *Le Temps* 7/2017, 4–6.

Keeling, Arn, and John Sandlos. 2017. 'Ghost Towns and Zombie Mines: The Historical Dimensions of Mine Abandonment, Reclamation, and Redevelopment in the Canadian

North', in Stephen Bocking and Brad Martin (eds), *Ice Blink: Navigating Northern Environmental History*. Calgary: University of Calgary, pp. 377–420.

Lagae, Johan, Jacob Sabakinu Kivulu and Luce Beeckmans. 2016. '"Pour Matadi la Question (de la Ségrégation) est encore plus Grave qu'Ailleurs": The Making and Shaping of a Congolese Port City During the Interwar Years', The Belgian Congo between the two World Wars Conference, 17–18 March. Brussels: Académie Royale des Sciences d'Outre-Mer.

Leech, Brian James. 2018. *The City That Ate Itself: Butte, Montana and Its Expanding Berkeley Pit*. Reno, NV: University of Nevada Press.

Mabogunje, Akin L. 1990. 'Urban Planning in the Post-Colonial State in Africa: A Research Overview', *African Studies Review* 33(3): 121–203.

Mansila Fu-Kiau, Shourit. 1988. 'Kolwezi: L'Émergence d'une Grande Ville Minière au Zaïre Méridonial', Ph.D. dissertation. Université de Lubumbashi.

Martinez-Fernandez, Cristina et al. 2012. 'Shrinking Cities: Urban Challenges of Globalization', *International Journal of Urban and Regional Research* 36(2): 213–25.

Mbembe, Achille. 2014. 'The Zero World: Materials and the Machine', in Sammy Baloji (ed.), *Mémoire/Kolwezi*. Brussels: Africalia & Stichting Kunstboek, pp. 73–79.

Mottoulle, Léopold. 1946. *Politique Sociale de l'Union Minière du Haut-Katanga pour sa Main d'Oeuvre Indigène et ses Résultats au Cours de Vingt Années d'Application*. Bruxelles, Institut Royal Colonial Belge.

Mukembe Mubedi, Marcellin. 2018. 'Kolwezi: du Peuplement Primitif à la Première Moitié du XXième Siècle (1700–1950)', *Les Cahier du CRESA: Revue d'études économiques & sociales appliquées* 51: 181–91.

Nash, June. 1979. *We Eat the Mines and the Mines Eat Us: Dependency and Exploitation in Bolivian Tin Mines*. New York: Columbia University Press.

Perrings, Charles. 1979. *Black Mineworkers in Central Africa: Industrial Strategies and the Evolution of an African Proletariat in the Copperbelt 1911–1914*. London: Heinemann.

Pugliese, Francesca. 2020. 'Mining Companies and Gender(ed) Policies: The Women of the Congolese Copperbelt, Past and Present', *Extractive Industries and Society*. Retrieved 26 September 2020 from https://doi.org/10.1016/j.exis.2020.08.006.

Rubbers, Benjamin. 2013. *Le Paternalisme en Question: Les Anciens Ouvriers de la Gécamines face à la Libéralisation du Secteur Minier Katangais (RDCongo)*. Paris: L'Harmattan.

Särkkä, Timo. 2016. 'The Lure of Katanga Copper: Tanganyika Concessions Limited and the Anatomy of Mining and Mine Exploration 1899–1906', *South African Historical Journal* 68(3): 318–41.

Udelsmann Rodrigues, Cristina, and Ana Paula Tavares. 2012. 'Angola's Planned and Unplanned Urban Growth: Diamond Mining Towns in the Lunda Provinces', *Journal of Contemporary African Studies* 30(4): 687–703.

— Chapter 5 —

Mining Life Cycles and Indigenous Land Dispossession in North America

A View from the American West

Paul J. White

Introduction

A notable consistency in the history of Western mining is how often mineral extraction has been conducted with limited regard to the rights and well-being of indigenous peoples. In North America, the westward expansion of Euro-American enterprise during the nineteenth century was intimately associated with the forced displacement of Native American and First Nation peoples from their land base. The general pattern remains evident today on the North American continent, as well as in interactions between Western mining enterprises and indigenous peoples globally (e.g. Ali 2003; Ballard and Banks 2003; Black 2018; Gedicks 2014; Kirsch 2014). Land loss has nevertheless taken many forms. Dispossession resulted from violent encounters as well as from the provisions of resource laws and global initiatives, and all of which have differed in their effects over time and space. This is to say that dispossession is perhaps best framed as an unfolding process rather than as an event with inevitable, evenly distributed impacts (e.g. Greer 2018; Meyer 1994; Saunt 2020).

Despite a robust and growing literature on the topic, historical treatments of interactions between miners and Native Americans occurring between the initial discovery rushes and the present day remain rare. An unintended consequence of this bifurcation in interest is that the discussion of mining heritage becomes similarly segregated into rush-era and post-rush era periods.

This chapter draws attention to the existence of long-term and continuing historical linkages between mining cycles and dispossession. I write this from the perspective of an archaeologist who has been involved in the historical assessment of US mining sites for more than two decades. A large portion of this work has involved applying models of heritage value to evaluate the significance of mining landscapes – in which a given site might be considered worthy of preservation according to its association with major patterns or events in US history, historical personages, or by being reflective of vernacular traditions. Such models have proven highly effective in recognizing the changing technological expressions of the industry (such as shifts to increasing economies of scale), but they remain largely internalist because they overlook the significance of continuing interactions between the industry and groups that mining has historically disenfranchised. This is no simple state of affairs. As explored here, part of this exclusion reflects historical circumstances by which Native Americans were physically evicted from regions deemed to have mineral potential, but this was rarely the end of the story. Attention to indigenous experiences, including from those who remained in mining districts, reveals just how difficult it is to cleanly parse histories along these lines. By way of example, I draw attention to the historical experiences of Timbisha Shoshone families in Death Valley, California, where land loss often occurred more than once, and where the consequences of mining activity extended beyond the transformations associated with mining booms.

Exploring National Historical Contexts

When European settlers set foot on the North American continent, their colonies, and later governments, sought to square idealized visions of a virgin land with the reality that their New World had long been peopled. Colonists' efforts to extinguish indigenous land title over the next few centuries fell under an increasingly varied set of strategies that ranged from skirmishes and declarations of outright war, to treaties with select indigenous groups and removal of others to marginal regions, to assimilationist programmes that endeavoured to 'kill the Indian … and save the man' through the imposition of Western values (Pratt 1892: 46). Today's existence of 574 federally-recognized tribes and sixty-three state-recognized tribes in the US draws attention to indigenous perseverance in the face of genocidal schemes. Such a tally nevertheless masks, among other circumstances, the harrowing experiences of historical trauma, the long histories of disruption to indigenous lifeways and the blatant and structural forms of racism that indigenous communities continue to face (Churchill 1997). Moreover, paths for obtaining legal recognition have proved to be convoluted. Many tribes remain unrecognized, and a lasting disjuncture also exists between legal status and the restitution of land title.

The legacy of Western mining in North America is deeply associated with these processes of dispossession, and in no small part because European interests in the North American continent from the outset included expectations of finding minerals for the taking (Mulholland 1981: 5–22; White 2017: 11–16). Fuelling this excitement was evidence that Native Americans were no strangers to mining. Beyond the common use of chert and obsidian, extensive trade networks had existed for turquoise, salt and native copper, and many indigenous groups also shared widespread beliefs that minerals like galena, ochre and quartz held spiritual power. Colonists used these various signs of indigenous usage to bolster allusions that the North American continent abounded in mineral wealth. Yet, they also tended to dismiss indigenous rights on the grounds that Native Americans did not smelt metals, and thus did not maximize the full economic potential of metal ores.

The discovery rushes of the nineteenth century saw this attitude manifested at an immense scale. The formation of mining districts seldom represented the first contact between indigenous and non-Native groups, but it nevertheless spurred interaction at a new level of intensity. Among the most well-documented cases occurred during the California Gold Rush (1849 to 1865). There, despite an early involvement of Native Americans in guiding prospectors to mineral outcrops and working mining claims, non-Natives quickly curtailed indigenous involvement.[1] By 1850, for instance, California courts had ruled Native American testimony invalid and also declared that vagrant Indians (as determined by whites) could be sold into indentured servitude (Johnston-Dodds 2002). Between 1845 and 1860, California's non-Native population rose from 10,000 to 350,000, while the territory's indigenous population dropped from an estimated 150,000 to 30,000 people as a consequence of violence, disease and the loss of a subsistence base (Hurtado 1998: 1, 100–24; Utley 1984: 2). Similar effects carried over to other regions as mineral discoveries came to be reported in other states and territories.

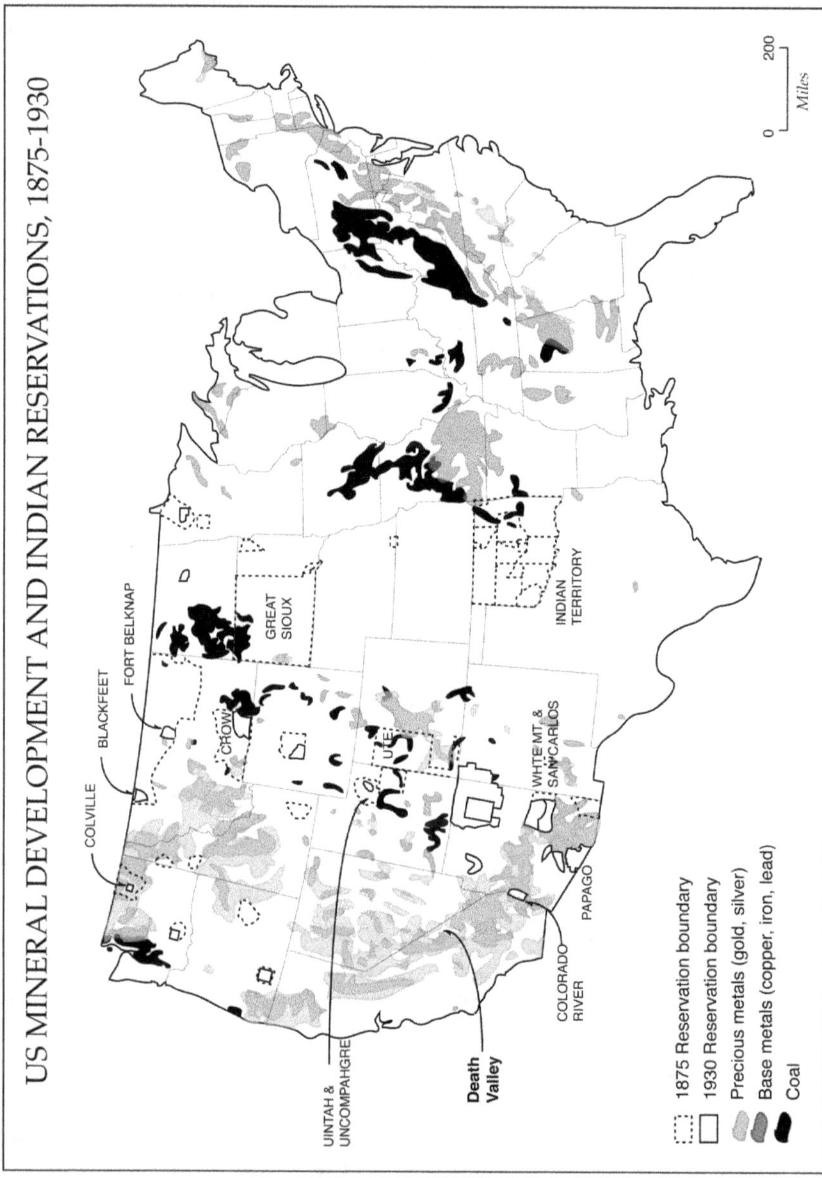

Figure 5.1. Map showing the tendency for Indian reservations to be sited away from mineral regions. Reservations symbolized by a dashed line represent boundaries in 1875; those symbolized by a solid line represent boundaries in 1930. Named reservations are those where the reduction of reservation lands was spurred in whole or in part by mineral discoveries. © Paul White.

Federal and state laws advanced the physical separation of Native Americans from mineral lands through three interrelated assumptions: that American Indians were to be assimilated to Euro-American values; that this cultural transformation would be accomplished best through the adoption of farming; and that mineral deposits on Indian lands should be developed whenever found (Black 2018; McDonnell 1991: 10–11; White 2008: 60–82). Thus, legislators established Indian reservations in areas considered most amenable to agriculture or ranching, and mineral discoveries in these areas could instigate the redrawing of reservation boundaries (Figure 5.1).

Even suspicions of mineral wealth could generate sufficient government pressure to trigger the forced removal of indigenous people from their homelands – two well-known examples being the Trail of Tears (1838), which ejected more than sixteen thousand Cherokee following the discovery of gold in Georgia, and the Navajo Long Walk (1864), in which rumours of mineral deposits in Arizona led to federal troops marching nine thousand Navajo from their lands (Ackerly 1998; Marks 1998; Williams 1993).

In some cases, mineral discoveries came to serve as the opening wedge for the total dissolution of reservation land – among the greatest loss in terms of acreage occurring on the heels of gold discoveries in the Black Hills of South Dakota in the 1870s, which precipitated the demise of the sixty-million-acre Great Sioux Reservation. A more typical experience involved the partial reduction of territory. Gold discoveries in Washington State, for instance, played a role in legislators reallocating the northern half of the Colville Indian Reservation to the public domain. Likewise, mineral discoveries along the Colorado River resulted in adjustments to the Colorado River Indian Reservation's boundaries in 1915 to exclude placer ground (Weber 2021). The US Congressional Record additionally reveals that several legislators sought to throw open all or at least most Indian reservations in the United States to prospecting, introducing at least forty such bills to Congress between 1885 and 1919, when it eventually passed as a rider to annual appropriations to the Bureau of Indian Affairs.[2] However, arguably the most surprising aspect is that the consistent removal of mineral lands occurred more through a smattering of individual laws than through any single policy. Moreover, the close timing between mineral discoveries and Acts opening specific reservations for mineral entry reveals how the inauguration of mining activity so often went hand in hand with the legal diminishment of indigenous land title.

The discovery rushes of the mid-nineteenth century also stimulated the development of laws intended to expedite the extraction of resources on lands already removed from the Native American estate. The General Mining Act (1866, 1872), which outlined procedures for staking mineral deposits on public lands, largely followed a policy of nongovernmental interference. Until a mining claim was patented, for instance, the government collected no taxes, and the legitimacy of a claim's worth was defendable by showing that one had annually expended effort on the property. As with other public land laws, the Mining Act was open to all US citizens, but this seemingly expansive provision excluded most Native Americans until the Citizenship Act of 1924 awarded US citizenship by default. While some public land policies, such as the Homestead Act, had undergone adjustments to allow Native American claimants on condition that they relinquish tribal affiliation, this did not apply to mineral lands. This is to say that legislators not only assumed that 'no Indian in the United States has ever been a miner', but they also crafted policies that ensured these assumptions would not be contradicted.[3]

One peculiar outcome of the legal separation of indigenous peoples from mineral lands has been a thematic replication of this divide in historical scholarship. While mining histories and Native American histories share considerable overlap when covering the discovery rush period, the existence of continuing and longer-term interactions seldom sees equivalent treatment. Broad overviews of Western mining, for instance, document the late nineteenth and early twentieth centuries in terms of the rise of wealthy capitalists, environmental transformations and industry shifts from small-scale operations to large-scale, capital-intensive and rationally organized enterprises (e.g. Curtis 2013; Gordon and Malone 1994: 177–218; Greever 1963; LeCain 2009; Lynch 2002; Paul 1947, 1963; Peterson 1977, 1991; Smith 1987; Spence 1970; Young 1970). These themes are manifested also by listings on the National Register of Historic Places (a national list of sites deemed worthy of preservation). There, sites that recognize indigenous mining heritage, such as copper and galena mines and flint and pipestone quarries, focus upon periods prior to European contact.[4] The Native American experience of mining is thus boxed in time as the earliest phase in a series of cycles.

Still, there is abundant evidence to indicate that the thematic separation of people, resources and time periods belies a messier reality. One starting point is that the legal distinction of mineral from nonmineral lands rested only upon an economic prediction about highest yield and not a statement about the activities possible for a given land tract. This was especially the case with arid portions of the American West, such as the Great Basin, where the ability for anyone to make a living so often hinged upon securing access to a reliable, good quality water supply. But the most obvious indicator of complexity is the ongoing presence of Native American communities in and around mining regions. While North America's most historically valuable mineral districts do not reside on Indian reservations, several Indian reservations are prized today as rich sources of oil, low-sulphur coal and uranium. Native American communities living on or near mining lands battle against the realities of pollution, resource depletion and inequitable leasing arrangements – the latter, according to scholars, ranking among the poorest agreements made between the mining industry and indigenous peoples worldwide (Ali 2003; Ballard and Banks 2003; Gedicks 1993, 2014; Snipp 1986a, 1986b; Voyles 2015).

Such considerations draw attention to the importance of investigating local settings, where more complicated and longer-term entanglements often took place. The remainder of this chapter presents the results of ethnohistorical and archaeological research exploring land loss in Death Valley, California, during a period when indigenous Timbisha Shoshone families found their seasonal camps overrun with mineral prospectors. My own research (White 2008) traced the land use patterns and land claiming efforts of three related Timbisha Shoshone families at water springs over the course of a century (1850 to 1950). Here, I draw attention to two fundamental points. The first is that close examination of archival records and material evidence reveals longer-term and more varied interactions between Timbisha Shoshone and miners than usually acknowledged. The second finding is that family histories, and specifically the tracking of claims and land use down through generations, add a vital approach to understanding how dispossession was tied to recurrent mining cycles.

Dispossession in Death Valley

During the nineteenth and twentieth centuries, most of the approximately three hundred members of the Timbisha Shoshone tribe resided in their homeland without federal recognition and without a federally recognized land base.[5] In common with Shoshonean and Paiute groups throughout the Great Basin, the Timbisha Shoshone have histories of occupation in the valley that date back at least several thousand years. For much of this time, subsistence practices involved living in extended family groups and moving seasonally between camps located on the valley floors during the winter and spring months and in the mountains during the summers. Ethnographic and oral historical evidence indicates that such strategies occurred without formal systems of private property. Non-Natives routinely misinterpreted the 'lack' of formalized property systems to indicate a limited connection to place. In contrast to these perceptions, Timbisha Shoshone learned stories that they, as a people, were born in the place and had obligations to maintaining the health of the land that ranged from the spiritual to pragmatic. Seasonal movements were far from being nomadic, for families had particular areas that they moved between, and a given family's range was known and generally respected by other families (Fowler et al. 1994).

Outsiders' perceptions of Shoshonean lifeways nevertheless served colonial interests in predictable and enduring ways. Shoshonean people were routinely cast as being 'primitive', one influential anthropologist deeming the culture to be 'gastric' (food-centred) in orientation and likening their social networks to 'a net without knots' (Steward 1938: 46; 1955: 58, 117). Such statements played powerful roles in dismissing indigenous land rights and cultural sovereignty (Blackhawk 1999; Ronaasen, Clemmer and Rudden 1999). The federal government's de facto classification of the Timbisha Shoshone as 'nonreservation Indians' legally dispossessed Timbisha Shoshone families of their land base without their knowledge. Historically, Timbisha Shoshone first experienced this state of affairs through the encroachment of mining interests. Miners trickled into Death Valley beginning in the 1850s and arrived in greater numbers during mineral rushes to the valley in the 1870s, 1890s, 1900s and 1930s (Lingenfelter 1986). To prospectors, these booms highlighted the existence of a wide range of potentially exploitable minerals. Although most deposits occurred in quantities that failed to justify sustained extraction beyond the mining of borax, talc and to a lesser extent gold, the recurrent interest in Death Valley's mineral potential made miners and prospectors a familiar presence.

The onset of mineral prospecting ushered in major economic transformations for the Timbisha Shoshone. Miners valued trees for firewood and construction, and this targeted the same resources that Timbisha Shoshone families depended on for subsistence activities, including nut and bean harvests. Water springs, where Timbisha Shoshone families had located seasonal camps, now became key places, too, for the siting of mining encampments. Miners hunted game to supplement their diets of canned food, and extensive grazing by their pack animals reduced seeds and grasses that Timbisha Shoshone had regularly used for dietary and medicinal purposes as well as for basketmaking. Absent of federal recognition, the Timbisha Shoshone had few options other than personal actions with which to negotiate these incursions. As with other Great Basin peoples, many Timbisha Shoshone took up wage work (or more often payment in the form of food). Typical 'Indian jobs' included timber cutting, hauling supplies to mining camps and leading prospectors to mineral outcrops (Knack 2001; Littlefield and Knack 1996). Some staking partnerships were amicable, but this did little to overcome the broader difficulties of adjusting to the sudden loss of a carefully maintained subsistence base, the inequitable and often paternalistic nature of labour arrangements, and the derision, if not outright dismissal, that Timbisha Shoshone people often received from non-Natives.[6]

Arguably Death Valley's most lasting historical irony is that the more this valley became familiar and accessible to outsiders, the more formidable it grew in myth as a place that tested the limits of human endurance. The perception of Death Valley as an inhospitable land propelled a belief among prospectors and promoters that the valley's sole redeeming quality had to reside in its mineral deposits; later visitors saw redemption more in terms of the valley's unique geographical attributes as the hottest, driest and lowest spot in North America. These factors became the prime reason behind establishing a 1.6-million-acre National Monument there in 1933, and for its later advancement, with land additions, to National Park status in 1994 (Lingenfelter 1986: 441–67; Rothman 2004).

The entrance of the National Park Service inaugurated another phase of dispossession for Timbisha Shoshone families. While the National Park Service were supportive of setting aside a forty-acre tract for an Indian colony near Monument headquarters in Furnace Creek, Monument staff were otherwise dismissive (Catton 2009: 26–29). A 1933 radio address by the National Park Service Director Horace Albright, for example, went so far as to contend that 'even the Indians never seriously claimed [Death Valley]' (quoted in Burnham 2000: 93). Monument correspondence also framed the presence of Timbisha Shoshone in Death Valley as a problem. Most Timbisha Shoshone activities, whether hunting, herding, collecting or moving seasonally, became prohibited or otherwise restricted because they conflicted with management goals to preserve and exhibit a natural environment for visitors. The National Park Service worked with the Bureau of Indian Affairs to develop 'Indian Village' – a permanent Indian settlement that they sited away from the mesquite groves where the Monument's visitor centre was located. Many Timbisha Shoshone nevertheless continued to move seasonally until the mid-1950s, when the Park began dismantling Indian Village over the course of a summer when tribal members were camping in the mountains. Since that time, the Timbisha Shoshone sought to maintain a permanent presence on the valley floor. Decades of continuing activism by tribal members eventually resulted in the tribe gaining federal recognition in 1983, and a land base in 2000 that includes acreage within and outside of Death Valley National Park (Catton 2009).

In contrast to agency attitudes to limit the Timbisha Shoshone presence, the National Park Service initially supported mining interests in both economic and symbolic ways. Park officials tapped mining companies during the first decade of the Monument's existence to assist with the development of roads to tourist overlooks (Figure 5.2).

In turn, Park officials granted some operators concessions to water springs. This friendly working relationship cooled considerably over time. In the 1970s, the National Park Service instituted a moratorium on mining until mineral claims could be appraised for their worth – and this action largely resulted in the eviction of mining operators from Death Valley.

Figure 5.2. Death Valley viewed from Aguereberry Point, a vista named after a local miner who helped construct the road to the overlook. © Paul White

Symbolically, however, the National Park Service continued to bolster mining's significance by recounting Death Valley's history in terms of Euro-American individuals and events. Several visitor destinations developed within the Monument elaborated upon the morbid place names such as the Funeral Mountains, Coffin Canyon and Starvation Gulch that miners had assigned to valley features. An overlook created in the Black Mountains was named Dante's View, a salt pan on the valley floor became the Devil's Golf Course, and a field of arrow weed shrubs, the Devil's Cornfield. Hell's Gate identified one of the routes into the Monument. The sites of early mining enterprises also became visitor destinations marked by memorials, as too did the graves of select prospectors who had entertained Monument visitors prior to their deaths and provided 'a great deal of color to the locality'.[7] This selective attention to the historical cultural landscape imposed an ideological form dispossession for the Timbisha Shoshone, for discussion of Death Valley's indigenous presence in the valley was largely relegated to a distant and precolonial past. Despite decades of entanglement, mining histories and Timbisha Shoshone histories were, and continue to be viewed as distinct interpretive islands.

Recovering Histories of Land Claiming

A very different picture of interaction emerges from documents preserved in county courthouses, Indian Agency files and National Park Service correspondence. Archival records reveal that the Timbisha Shoshone pursued formal recognition of their land rights for several decades through numerous legal channels. At the time of the establishment of Indian Village at Furnace Creek in 1937, for instance, Shoshonean claimants had applied for and received two 80-acre Indian Homesteads in 1885, filed three water appropriation notices in 1896 and made verbal inquiries about several more, been awarded a 1,560-acre rancheria in 1928, obtained three Indian Allotments (two 160-acre allotments in 1927 and a 40-acre award in 1936), and been involved in staking nearly three hundred mining claims in the region (Crum 1998, 2002; White 2008) (Figure 5.3).

Beyond this tally of land acquisitions, a closer reading of these filings highlights three themes: that land and resource applications dovetailed with the arrival of mining interests, that knowledge about land claiming passed through kin networks, and that Timbisha Shoshone experienced problems in materially benefiting from legal gains.

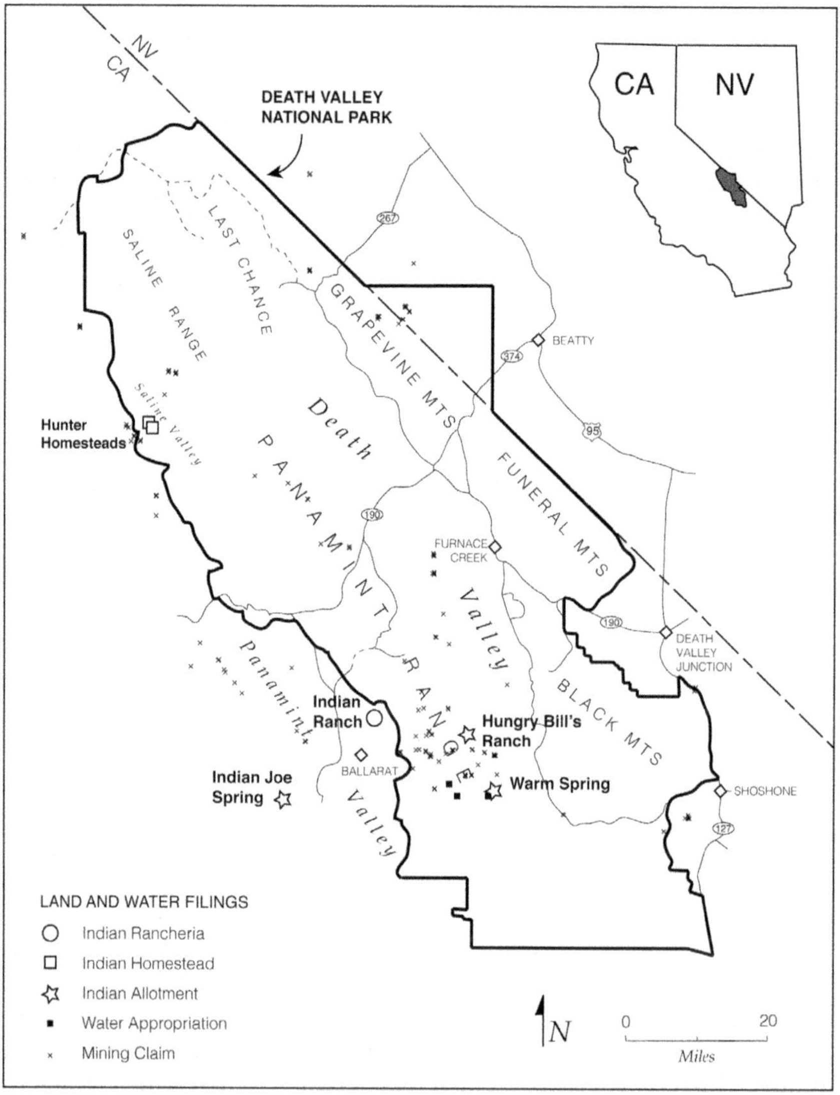

Figure 5.3. Official Timbisha Shoshone claiming activities from 1870 to 1950, as documented in county records and Indian Agency files. © Paul White.

Historic Timbisha Shoshone land claiming occurred mostly in response to miners encroaching upon key areas. A telling example of this tendency occurred at Warm Spring, a permanent good quality water source located within the southern Panamint Mountains and used as a winter camp by generations of the Tom family. Panamint Tom filed an annual water appropriation to the spring in 1896, the same year that newspapers announced that prospectors had discovered promising mineral outcrops in this section of the Panamints. Under US resource law, a water appropriation gave claimants legal authority to use water flowing on the public domain to the exclusion of other users. A typical notice described the claimant's intentions to take a stipulated volume of water from a given water spring and to convey it by means of a pipe of a specified diameter for so many feet. But the notices filed by Panamint Tom and one of his sons, Indian Harry, differed markedly in intent. Indian Harry's single filing claimed all the water at Anvil spring as well as another spring located a half mile distant and all the ground between them. Panamint Tom added in a witness statement that 'I the undersigned have lived here 43 years and claim all of these springs now and others'. Panamint Tom's own water appropriation notices to water springs expressed similar sentiments. At the site of his summer camp, which he named 'Panamint Toms Spring Mine', he claimed all the water and 'all timber from summit to said water'. A second filing claimed Warm Spring as well as Clear Water Spring, water sources also located within his family's traditional range but situated nine miles distant from each other.[8] The legality of these filings may have been questionable, but the intention to announce ownership is clear.

Water appropriation notices needed to be filed each year to remain valid, but this step seems unlikely to have been pursued. In 1910, Panamint Tom inquired about filing for a homestead under the 1875 Indian Homestead Act after a group of miners had filed a water appropriation to Warm Spring. Tom's application was apparently lost, but in 1916, Hungry Bill (Panamint Tom's brother) made another inquiry for a homestead at Warm Spring on behalf of Tom's son, Bob Thompson. This second attempt also proved ineffective because administrators became confused as to whether Panamint Tom and Hungry Bill (also known as Panamint Bill) were the same person. Mining activity again returned to Warm Spring in 1929. Bob Thompson first attempted to lease the water rights to miners who had set up camp there, but they reneged on the deal. This situation provoked Thompson to file for an Indian Allotment under the 1887 Dawes Act. At the time of his application in 1934, then, the

Tom family had tried at least four previous times over the past four decades to assert rights to the same place, and with each occasion triggered by miners' encroachment (Crum 1998; White 2006).

The timing of other Timbisha Shoshone claims suggests a similar association. Hungry Bill, for example, filed for an Indian Allotment in Johnson Canyon in 1907 shortly after prospectors expressed interest in outcrops near the site of his family's summer camp. The Hunter family's Indian Homesteads in Saline Valley were filed in the 1880s, shortly after a mining company had applied to use water from the canyon where the Hunter family summered. A later water appropriation filed on the Hunter family's behalf provided them with legal leverage for negotiating water use with the company (the arrangement allowed the borax mining company to take half of the water for fifty dollars a year and an assurance to employ family members at the borax works).[9] The application for an Indian rancheria in Panamint Valley came after flash floods had caused George Hanson's family to temporarily abandon their winter camp. A supervising engineer inspecting the property on behalf of the Bureau of Indian Affairs feared that the temporary absence of the Hansons would open the door to miners and other interested parties appropriating the water and other improvements that the family had made over the generations.[10]

The involvement of forty Timbisha Shoshone individuals in staking mining claims may also have been prompted by desires to have some say in the negotiation of resources. Among the approximately three hundred mining claims staked over the course of four decades, it is notable that nearly all Timbisha Shoshone claimants staked properties within their traditional family ranges. The primary exception was Bob Thompson, who Elders recall as being something of a miner, and who staked claims throughout Death Valley and neighbouring areas. Even here, however, Thompson's claiming activities indicate that he returned most often to the southern portion of the Panamint Mountains, the place where he was born and raised (White 2006: 8–9). As with water appropriations, most mining claim filings were one-off occurrences and not followed up with annual proof of labour forms necessary to keep a claim valid.

Timbisha Shoshone claiming patterns additionally suggest that knowledge about threats and claiming strategies travelled through kin networks. Although most Timbisha Shoshone partnered with non-Natives in filing mining claims, claim records also indicate Timbisha Shoshone filed mining claims by themselves and in exclusive partnerships with their kin.[11] Bob Thompson, for example, staked several mining claims in partnership with his son, stepson and two cousins. One of his cousins, Susie Wilson, staked two mining claims with her 7-year-old son – with both claims located in the vicinity of where her family summered.

An additional indication of kin networks is seen through the patterns of land applications filed at the same time. Panamint Tom, for instance, travelled to the county seat to file his water appropriation notices along with those belonging to his son; the two applications by the Hunter family for Indian Homesteads in the Saline Valley had been filed simultaneously; and Hungry Bill's filing for an Indian Allotment occurred at the same time as his cousin also filed for an Indian Allotment in the Searles Valley, thirty miles distant. Thus, while legal filings necessarily conferred rights to individual claimants, practically all claimants were related – as siblings, nephews, cousins and in-laws. Death Valley National Monument staff seemed aware of the broader threat that such land inquiries posed. In the 1930s, Monument staff discouraged Susie Wilson's husband from obtaining use rights to a set of water springs in the Panamints that he hoped to use for ranging horses. More particularly, they were concerned that allowing him to have water rights would open the door to 'an Indian claiming every good water supply in the Valley', and so they opted instead to allow a water rights application by a mining company.[12]

Timbisha Shoshone land and water filings reveal repeated ways that mining activity was an active concern, but case histories also highlight recurrent problems that Timbisha Shoshone families faced in perfecting land title. A key problem rested on a structural inequality in the efficacy of public land law and federal Indian policy: namely, applying for Indian Homesteads and Indian Allotments involved more bureaucratic steps than filing for mining claims and water appropriations. To secure ownership to a mining claim, a prospector staked a property, marked the corners and then filed a copy of the claim notice at the county recorder's office within ninety days. For a nonreservation Indian to file for an Indian Allotment, however, the claimant had to first seek a Justice of the Peace, who would confirm that the claimant was Native American and file the paperwork on the claimant's behalf. The county recorder's office sent the application to the nearest Indian agency, who might send out an inspector to validate the claim and claimant. This being accomplished, the application was next mailed to the central office of the Bureau of Indian Affairs, who might send out a Special Agent to investigate the property if the claim had resulted from a dispute. Once the Bureau of Indian Affairs had approved the application, the General Land Office sent out a surveyor to record the property, with the final certificate then delivered back to the nearest Indian agency.

At best, this process could take two years, and it could also take much longer. Hungry Bill and Joe Peterson filed for allotments in 1907, but their applications stagnated within the offices of the Bureau of Indian Affairs and General Land Office. Misplaced several times by branch offices, these applications ultimately passed through four land offices and nine Indian agencies (Figure 5.4).

When finally awarded in 1927, Hungry Bill had been dead for a decade, and the prospecting boom in Johnson Canyon that had prompted the filing had long passed. Administrative red tape had thus allowed miners to pursue their interests without constraint (White 2008: 191–94).

Beyond the delays caused through the structural inequalities of land laws, it is also apparent that property title fell short of capturing use. The surveyed boundaries of Hungry Bill's allotment, for

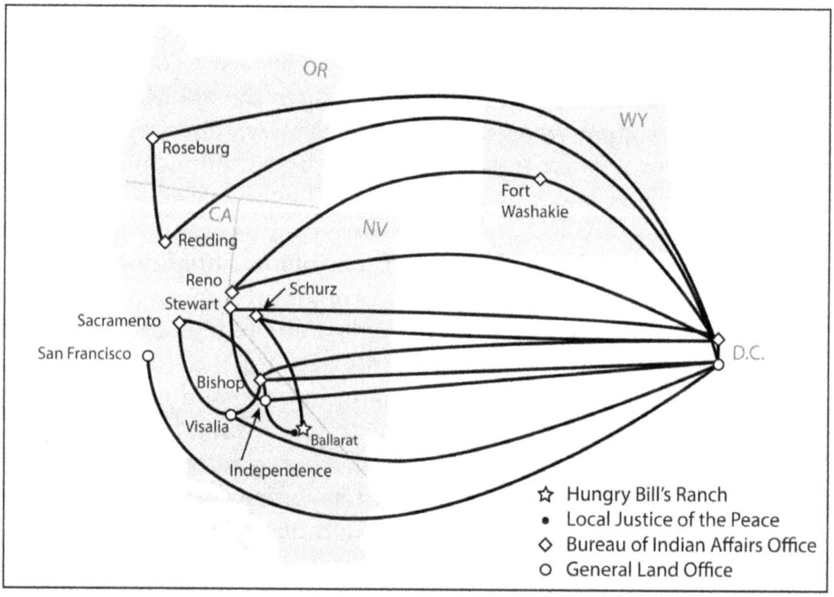

Figure 5.4. The geography of red tape. Correspondence concerning Hungry Bill's allotment filing, 1907–27. If all had gone according to procedure, this application would have taken approximately two years and passed through offices in Independence, Bishop and Washington, DC, after being authorized by a Justice of the Peace in Ballarat. Instead, bureaucratic misfilings drew out the process for twenty years. Miners at Hungry Bill's Ranch, by contrast, could stake mining claims themselves and simply had to file them in Independence within 90 days. © Paul White.

example, show significant discrepancies between lands used and the property awarded (Figure 5.5).

As indicated by an archaeological survey, the Hungry Bill family intensively utilized a three-mile stretch of narrow Johnson Canyon, with material remains including campsites, two horse corrals, several fields bordered by stone walls and wooden fences, and two orchards (Pearson 2005). Hungry Bill's granddaughter, who lived at the ranch until her early teens, also recalls regular visits to pinyon pine stands (whose nuts supplied a key summer food staple) located further up canyon as well as to the south.[13] The boundaries of Hungry Bill's Indian Allotment excluded practically all of these features, although it did include the main water spring at its southwest corner.

Bob Thompson encountered a similar problem through different circumstances. He received his Indian Allotment within two years of applying, but the miners occupying Warm Spring vehemently disputed the validity of his claim. The miners had staked a gold mining claim directly over Warm Spring around the same time that Thompson filed his application, and they had also been dutiful in filing proof of labour forms needed to validate the claim each year.

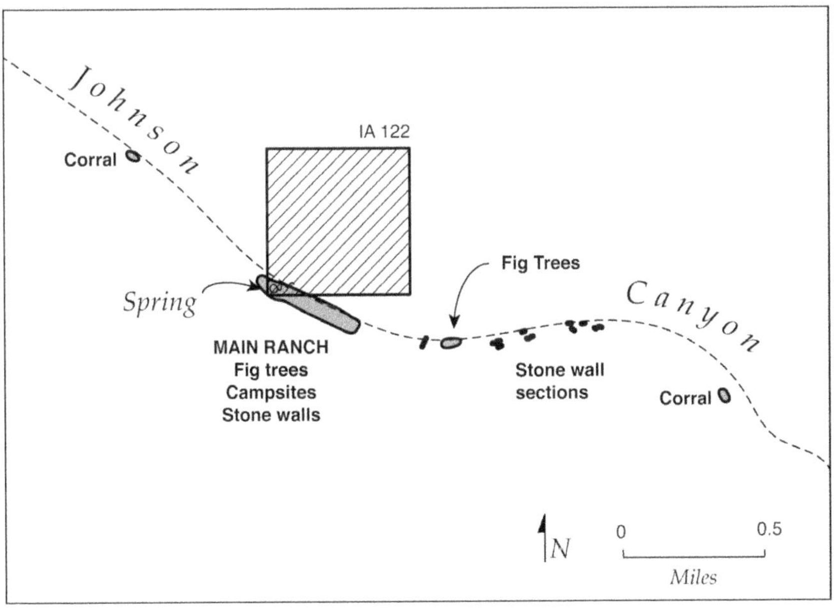

Figure 5.5. Hungry Bill's Indian Allotment (IA 122) plotted in relation to the distribution of historic cultural features associated with his family's use of Johnson Canyon. © Paul White.

The ownership dispute continued until 1941, when a court ruling upheld the validity of the Indian Allotment except for areas claimed by mining interests. A survey of claim boundaries and cultural features at Warm Spring reveals what a Pyrrhic victory the court ruling meant for Thompson (Figure 5.6).

The mining claim and mill site staked by the miners took in Warm Spring proper as well as its watercourse, the site of Thompson's former cabin (that miners had since scrapped for firewood), and the orchard that he and his father had tended for decades. An archaeological survey of the mining work ultimately conducted on the property indicates that it could likely have been accomplished over the course of a couple of years (White 2006: 16–17). Outside of the mining claims, Thompson's property consisted primarily of gravel and rock, with little use for either subsistence or mineral exploration. Indeed, following the court victory, the miners ended gold prospecting and shifted their attention to the mining of nearby talc

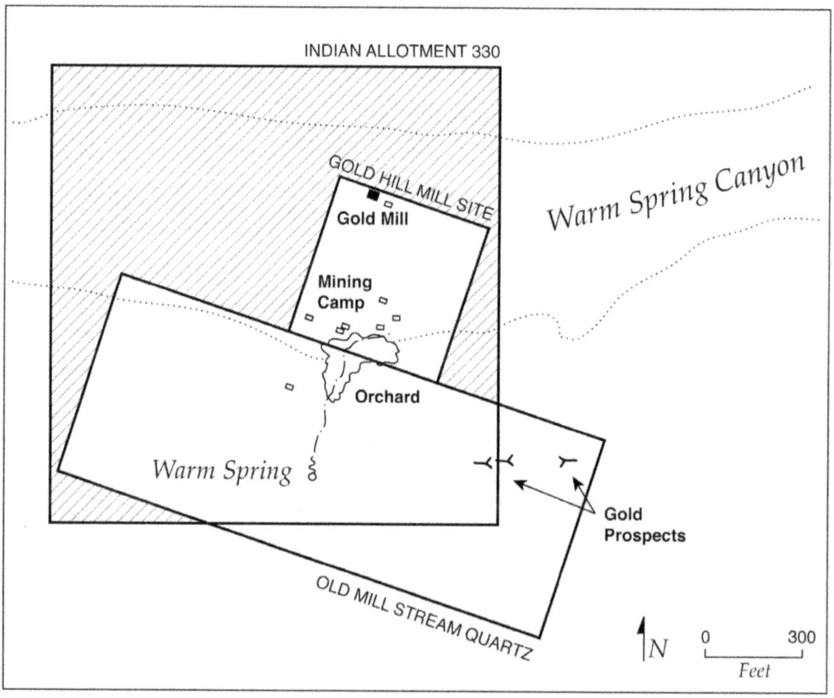

Figure 5.6. Contested land claims at Warm Spring, showing location of water spring and orchard within the boundaries of the mining claims. The shaded area indicates the portion of the allotment that Bob Thompson still retained ownership of after the conclusion of the court trial. © Paul White.

claims, while keeping their camp at Warm Spring. Here, the Mining Act had proven to be of greatest value in providing an expedient way for miners to claim a prized water source.

Such occurrences prompted at least one Timbisha Shoshone family to apply a different strategy for asserting land rights. During the 1920s and 1930s, the Hansons faced considerable pressure from miners and ranchers who sought to limit the range of their ranching activities. This included a miner fencing off an important spring that the Hanson family used when moving goats to summer pastures, and ranchers setting fire to one of the Hansons' camps. With efforts to legally secure water springs becoming increasingly difficult, the Hansons seem to have resorted to extra-legal tactics. Scattered throughout the Argus and Coso Ranges, located west of the Panamint Mountains, are boulders carved with 'LH' (for Logan Hanson) as well as 'M' with an underscore (Figure 5.7).

Shown a couple of these sites in 2006, a descendant of the Hanson family noted: 'My family did not believe in registering their brands. They were busy. It was our land. Our brand was an M with a bar

Figure 5.7. The brands of Mike Hanson and Logan Hanson (note backwards '9' in the 1934 date) carved into a rock at Tennessee Spring in the Argus Range. © Paul White.

over it. Sometimes we got it backwards and put an upside-down M with a bar under it. It meant Mike Hanson, my grandfather's cows and horses' (quoted in Ruby and Johnson 2007: 39).

The Hanson family also informally claimed a water spring that they had long used south of their rancheria after a miner had filed a placer claim on the spring in 1939 with the idea of turning it into a tourist resort. Nothing came from the resort plans, and the Hansons responded the following year by fencing a paddock at the spring and keeping a few horses pastured there (Figure 5.8). This plot was never formally applied for, but the barbed wire fence proved effective in dissuading others from further encroachment.[14]

Broadening the Interpretive Frame

Viewed only at the coarsest scale of resolution, interactions between miners and Timbisha Shoshone families occurring between 1850 and 1950 might first appear negligible. Records of violent conflict are comparatively sparse, the federal government extended no protections to the Timbisha Shoshone as a tribal group, and only a few Timbisha

Figure 5.8. Mesquite-pole fence constructed by the Hanson family in 1940 at a water spring south of their rancheria. © Paul White.

Shoshone individuals acquired land title. But a closer examination of archival records and material evidence reveals just how far simplistic interpretations stray from historical reality. Throughout this period, the Timbisha Shoshone made repeated attempts to assert their inalienable rights, and this occurred especially when miners encroached on water springs. Timbisha Shoshone efforts to claim ownership were varied in approach, and their inquiries were also filtered through an imperfect system of land laws that favoured permanent over seasonal use and public land laws over Indian policy. Irrespective of the outcome of Timbisha Shoshone inquiries, a tight linkage continued to exist between mining cycles and land dispossession. Timbisha Shoshone experienced dispossession not as a one-time event but as a recurrent process. They endured land loss whenever miners came into Death Valley, and sometimes in the exact same places.

The circumstances seen in Death Valley resonate with indigenous experiences elsewhere in the American West. Records for Indian agencies that administered Shoshonean and Paiute groups in the Great Basin, for example, reveal numerous inquiries about land rights and the difficulties that indigenous claimants experienced in securing them (Crum 2002; Fowler, Dufort and Rusco 1995; Knack 2001; White 2008). Among the historical correspondence are repeated indications of the pressure that non-Native interests placed on water resources, instances by which mining companies failed to honour leasing arrangements, and the misuse of the Mining Act to claim water springs and other valued nonmineral resources. These situations were experienced also by Native American communities who had been afforded federal recognition and a land base. In southern Arizona, for example, the Tohono O'odham reported abuses of the Mining Act as part of an unsuccessful protest against their reservation's reopening to mineral entry in 1934. There, and within months of being opened to entry, non-Natives had used the Mining Act's lenient provisions to claim several important water springs on the reservation, and upon which these claimants subsequently developed gas stations, whiskey stills, and residences (US Senate Committee on Indian Affairs 1934: 75–78).

One of the takeaways is the need to complicate interpretations of mining heritage by recognizing long-term and continuing entanglements between industry practitioners and indigenous groups. A key pattern in US history is that fundamental resource conflicts – including disagreements over what is an exploitable 'resource' – remain unresolved. The arrival of non-Native interests in Death Valley in the 1850s and the reclassification of the region as a National Park in the 1990s essentially bookend the extractive timescape. Even so, gold mining continues on the outskirts of Death Valley, and it remains a leading industry in the Great Basin, where operations presently contribute to just under 10 percent of the world's production. Here, as well as worldwide, the increasing scale of mining operations has opened new arenas for old conflicts. Today, disputes between miners and indigenous groups (including the Timbisha Shoshone) frequently escalate beyond the boundaries of individual water springs to concern entire water aquifers, and beyond the scale of individuals and families to impacts affecting multiple tribal groups. The discomforting historical consistency is that the mining industry continues to perpetuate unequal relations with indigenous groups nationally and worldwide. This situation recurs at the national level because industry practitioners continue to operate according to resource laws that originated in colonial circumstances to serve colonizing ends. It is promulgated similarly at the global level through neoliberal policies that promote the privatization of public assets beyond a nation's borders.

Acknowledgements

Thanks to the Timbisha Shoshone tribe for sharing their knowledge about their homeland, and to Antonio Maria Pusceddu and Filippo Zerilli for organizing the 2019 Global Life of Mines event.

Paul J. White is an Associate Professor of Geography at the University of Nevada, Reno, with 25 years of experience in the archaeological documentation of mining sites in North America. His research interests centre on the social, technological and environmental transformations associated with mining. He has worked also as a consultant for the U.S. Environmental Protection Agency and conducted safety hazard assessments on abandoned mining lands. His articles on mining have appeared in *Historical Archaeology, IA: The Journal of the Society for Industrial Archeology*, and *The Mining History Journal*. His book, *The Archaeology of American Mining* (2017), which synthesizes 50 years of archaeological scholarship, received the 2019 Mining History Association's Clark Spence Award.

Notes

1. Indeed, as historian James Rawls notes, Native American labourers excavating the tailrace for John Sutter's sawmill were the likely discoverers of gold nuggets that prompted the California Gold Rush (Rawls 1976: 28–30).
2. Based on the author's examination of bills and debates concerning mineral deposits on Indian lands, as published in the US Congressional Record, vols. 62 to 86 (1862 to 1941). Regarding the redrawing of reservation lands, this situation applied to 'Indian Territory' that became Oklahoma, the Blackfeet, Crow and Fort Belknap reservations in Montana, the Colville reservation in Washington, the Great Sioux reservation in South Dakota, the Uintah and Uncompahgre reservations in Utah, the Ute reservation in Colorado, and the Colorado River, Papago and White Mountain-San Carlos reservations in Arizona. Additionally, mineral lands identified on Indian reservations undergoing allotment (a process that partitioned reservation land into units of private property) were automatically transferred to the public domain.
3. The Fourteenth Amendment to the US Constitution (1868) defined citizenship as 'all persons born or naturalized in the United States'. US courts, however, variously interpreted Native American tribes as being 'foreign nations', 'domestic dependent nations' and 'wards', all of which excluded citizenship. One example of a public land law adjusted for Native American claimants was the passage of the Indian Homestead Act (1875) that supplemented the Homestead Act (1862) to allow Native American entry under the condition that claimants relinquish tribal affiliation. The 1924 Citizenship Act, by contrast, assigned Native Americans US citizenship irrespective of this condition. 'No Indian …' quote from Senator Albert Fall. Congressional Record, 65th Cong., 2d sess., 1918 (56): 7935.
4. This includes the Mines of Spain in Iowa for galena, the Minong Mine Historic District on Isle Royale, Michigan, which preserves prehistoric copper pits, and quarry sites at Pipestone National Monument in Nebraska and the Alibates Flint Quarry in Texas. See Noble and Spude, 'Guidelines for Identifying, and National Register database'. Retrieved from https://www.nps.gov/subjects/nationalregister/database-research.htm.
5. Although the signing of the Treaty of Ruby Valley in 1863 recognized Shoshonean land title to a vast area of the Great Basin including Death Valley, the Timbisha Shoshone were not signatories on the treaty and were unlikely aware of its existence until well after the fact. The Treaty ruled that these lands (approximately 24 million acres) could be prospected and worked for gold and silver, among other minerals, without interference. See Crum (1994: 25–26).
6. Several Timbisha Shoshone men are identifiable in records by the names either adopted or assigned by their employers (for instance, Indian George, Indian Joe, Indian Pete and Indian Tom). As occurring elsewhere throughout the Great Basin, some Timbisha Shoshone women also entered into domestic partnerships with non-Native miners. Many such marriages were never formalized and ended when the miners left. The labels 'squaw man' and 'half breed' highlight the disdain to which outsiders viewed such unions, and indigenous peoples more generally.
7. Col. John R. White to National Park Service Director, 24 February 1931. Situation Reports, 1931–38, D30317, Death Valley National Park Archives, Death Valley, CA. The particular miners in question included John Lemoigne, Shorty Harris, James Dayton and Death Valley Scotty.
8. Water appropriation notices recorded 10 August 1896, Inyo County; Land, Water, and Mining Claims, Book I: 410, 413. Inyo County Courthouse, Independence, CA.

9. Fowler, Dufort and Rusco (1995: 85); Water appropriation notice recorded 2 January 1890, *Inyo County; Land, Water, and Mining Claims*, Book F: 38, Inyo County Courthouse, Independence, CA.
10. Herbert V. Clotts to Commissioner of Indian Affairs, 4 June 1927. RG75 Sacramento, Tribal Group Files 1915–72, Box 35, Indian Ranch Owens Valley, National Archives San Bruno, CA.
11. Based upon author's investigation that examined mining claim notices for Inyo County, California and Nye and Esmeralda counties, Nevada, in relation to names listed on tribal rolls.
12. T. Ray Goodwin to Col. John R. White, 17 July 1934. L30 Landuse Indian Village 1933–39, D28140a-d, Death Valley National Park Archives, Death Valley, CA.
13. Agnes Sudway, (Elder) Timbisha Shoshone Tribe, personal communication May 2006.
14. Frank Parcher to Don Foster, February 1940, RG 75 Sacramento, Tribal Group Files 1915–72, Box 35, Indian Ranch Owens Valley, National Archives San Bruno, CA.

References

Ackerly, Neal W. 1998. 'A Navajo Diaspora: The Long Walk to Hwéeldi'. Retrieved 21 September 2021 from https://bloodhound.tripod.com/longwalk.htm.

Ali, Saleem. 2003. *Mining, the Environment, and Indigenous Development Conflicts*. Tucson, AZ: University of Arizona Press.

Ballard, Chris, and Glenn Banks. 2003. 'Resource Wars: The Anthropology of Mining', *Annual Review of Anthropology* 32: 287–313.

Black, Megan. 2018. *The Global Interior: Mineral Frontiers and American Power*. Cambridge: Harvard University Press.

Blackhawk, Ned. 1999. 'Julian Steward and the Politics of Representation', in Richard O. Clemmer, L. Daniel Myers and Mary Elizabeth Rudden (eds), *Julian Steward and the Great Basin: The Making of an Anthropologist*. Salt Lake City, UT: University of Utah Press, pp. 203–18.

Burnham, Philip. 2000. *Indian Country, God's Country: Native Americans and the National Parks*. Washington, DC: Island Press.

Catton, Theodore. 2009. 'To Make a Better Nation: An Administrative History of the Timbisha Shoshone Homeland Act'. Report on file at Death Valley National Park Archives, Cow Creek, CA.

Churchill, Ward. 1997. *A Little Matter of Genocide: Holocaust and Denial in the Americas, 1492 to the Present*. San Francisco: City Lights.

Crum, Steven. 1994. *Po'i Pentum Tammen Kimmappeh, The Road on Which We Came: A History of the Western Shoshone*. Salt Lake City, UT: University of Utah Press.

———. 1998. 'A Tripartite State of Affairs: The Timbisha Shoshone Tribe, the National Park Service, and the Bureau of Indian Affairs, 1933–1994', *American Indian Culture and Research Journal* 22(1): 117–36.

———. 2002. 'Pretending They Didn't Exist: The Timbisha Shoshone Tribe of Death Valley, California, and the Death Valley National Monument up to 1933', *Southern California Quarterly* 84(3–4): 223–40.

Curtis, Kent A. 2013. *Gambling on Ore: The Nature of Metal Mining in the United States, 1860–1910*. Boulder: University of Colorado Press.

Fowler, Catherine S., Molly Dufort, Mary Rusco, and the Historic Preservation Committee, Timbisha Tribe. 1994. 'Residence Without Reservation: Ethnographic Overview and

Traditional Land Use Study, Timbisha Shoshone, Death Valley National Monument, California'. Report on file at Timbisha Shoshone Tribal Office, Death Valley, CA.
Fowler, Catherine S., Molly Dufort and Mary Rusco. 1995. 'Timbisha Shoshone Tribe's Land Acquisition Program: Anthropological Data on Twelve Study Areas'. Report on file at Timbisha Shoshone Tribal Office, Bishop, CA.
Gedicks, Al. 1993. *The New Resource Wars: Native and Environmental Struggles Against Multinational Corporations*. Boston: South End Press.
______. 2014. *Dirty Gold: Indigenous Alliances to End Global Resource Colonialism*. Cambridge, MA: South End Press.
Gordon, Robert, and Patrick Malone. 1994. *The Texture of Industry: An Archaeological View of the Industrialization of North America*. New York: Oxford University Press.
Greer, Allan. 2018. *Property and Dispossession: Natives, Empires and Land in Early Modern North America*. Cambridge: Cambridge University Press.
Greever, William S. 1963. *The Bonanza West: The Story of the Western Mining Rushes, 1848–1900*. Norman, OK: University of Oklahoma Press.
Hurtado, Albert. 1998. *Indian Survival on the California Frontier*. New Haven, CT: Yale University Press.
Johnston-Dodds, Kimberly. 2002. 'Early California Laws and Policies Related to California Indians'. California Research Bureau report 02-014. Retrieved 21 September 2021 from https://cawaterlibrary.net/wp-content/uploads/2017/05/Early-Laws-Pertaining-to-Tribes.pdf.
Kirsch, Stuart. 2014. *Mining Capitalism: The Relationship Between Corporations and Their Critics*. Berkeley, CA: University of California Press.
Knack, Martha. 2001. *Boundaries Between: The Southern Paiutes, 1775–1995*. Lincoln, NE: University of Nebraska Press.
LeCain, Timothy J. 2009. *Mass Destruction: The Men and the Giant Mines that Wired America and Scarred the Planet*. New Jersey: Rutgers University Press.
Lingenfelter, Richard. 1986. *Death Valley and the Amargosa: A Land of Illusion*. Berkeley, CA: University of California Press.
Littlefield, Alice, and Martha Knack (eds). 1996. *Native Americans and Wage Labor: Ethnohistorical Perspectives*. Norman, OK: University of Oklahoma Press.
Lynch, Martin. 2002. *Mining in World History*. London: Reaktion Books.
Marks, Paula Mitchell. 1998. *In a Barren Land: American Indian Dispossession and Survival*. New York: William Morrow and Company.
McDonnell, Janet A. 1991. *The Dispossession of the American Indian, 1887–1934*. Bloomington, IN: Indiana University Press.
Meyer, Melissa L. 1994. *The White Earth Tragedy: Ethnicity and Dispossession at a Minnesota Anishinaabe Reservation*. Lincoln, NE: University of Nebraska Press.
Mulholland, James A. 1981. *A History of Metals in Colonial America*. Tuscaloosa, AL: University of Alabama Press.
Noble, Bruce, and Robert Spude. 1997. *Guidelines for Identifying, Evaluating, and Registering Historic Mining Sites*, National Register Bulletin 42, rev. ed. Washington, DC: National Park Service.
Paul, Rodman W. 1947. *California Gold: The Beginning of Mining in the Far West*. Lincoln, NE: University of Nebraska Press.
______. 1963. *Mining Frontiers of the Far West, 1848–1880*. New York: Holt, Rinehart and Winston.
Pearson, Nancy. 2005. 'Rock Walls and Wooden Fence Posts: Archaeological Inventory and Ethnohistorical Research in Johnson Canyon, Death Valley National Park, Inyo County, California'. Tucson: National Park Service, Western Archaeological and Conservation Center.

Peterson, Richard H. 1977. *The Bonanza Kings: The Social Origins and Business Behavior of Western Mining Entrepreneurs, 1870–1900*. Lincoln, NE: University of Nebraska Press.

———. 1991. *Bonanza Rich: The Lifestyles of the Western Mining Entrepreneurs*. Moscow, ID: University of Idaho Press.

Pratt, Richard H. 1892. 'The Advantages of Mingling Indians with Whites', in *Proceedings of the National Conference of Charities and Correction at the Nineteenth Annual Session Held in Denver, Col., June 23–29, 1892*, ed. Isabel C. Barrows, 45–59 (Boston: George H. Ellis), reprinted in *Americanizing the American Indians: Writings by the "Friends of the Indian" 1880–1900*, ed. Francis Paul Prucha, 260–71. Cambridge, MA: Harvard University Press.

Rawls, James. 1976. 'Gold Diggers: Indian Miners in the California Gold Rush', *California Historical Quarterly* 55(1): 28–45.

Ronaasen, Sheree, Richard O. Clemmer and Mary Elizabeth Rudden. 1999. 'Rethinking Cultural Ecology, Multilinear Evolution, and Expert Witnesses: Julian Steward and the Indian Claims Commission', in Richard O. Clemmer, L. Daniel Myers and Mary Elizabeth Rudden (eds), *Julian Steward and the Great Basin: The Making of an Anthropologist*. Salt Lake City, UT: University of Utah Press, pp. 170–202.

Rothman, Hal K. 2004. 'To Ride Alone in a Forever Unpossessed Country: An Administrative History of Death Valley National Park'. Report on file at Death Valley National Park Archives, Cow Creek, CA.

Ruby, Allika, and Lynn Johnson. 2007. 'A Section 110 Cultural Resources Inventory and Overview of 4,950 Acres West of the Argus Range and East of Darwin Wash, Naval Air Weapons Station, China Lake, Inyo County, California'. Far Western Archaeological Research Group, 1940. On file at China Lake Naval Air Weapons Station.

Saunt, Claudio. 2020. *Unworthy Republic: The Dispossession of Native Americans and the Road to Indian Territory*. New York: W. W. Norton.

Smith, Duane A. 1987. *Mining America: The Industry and the Environment, 1800–1980*. Lawrence, KA: University Press of Kansas.

Snipp, C. Matthew. 1986a. 'The Changing Political and Economic Status of the American Indians: From Captive Nations to Internal Colonies', *American Journal of Economics and Sociology* 45(2): 142–57.

———. 1986b. 'American Indians and Natural Resources: Indigenous Peoples' Land, Now Sought After, Has Produced New Indian-White Problems', *American Journal of Economics and Sociology* 45(4): 457–74.

Spence, Clark C. 1970. *Mining Engineers and the American West: The Lace Boot Brigade, 1849–1933*. New Haven, CT: Yale University Press.

Steward, Julian H. 1938. 'Basin-Plateau Aboriginal Sociopolitical Groups', *Bureau of American Ethnology Bulletin* 120. Washington, DC: GPO.

———. 1955. *Theory of Culture Change: The Methodology of Multilinear Evolution*. Urbana, IL: University of Illinois Press.

US Senate Committee on Indian Affairs. 1934. *Hearing Before the Committee on Indian Affairs, United States Senate, 73rd Congress, 2nd Session on S. J. Res. 95; A Joint Resolution Restoring Lands of the Papago Indian Reservation in Arizona to Exploration and Location Under the Public Land Mining Laws, April 24, 1934*. Washington, DC: GPO.

Utley, Robert. 1984. *The Indian Frontier of the American West, 1846–1890*. Albuquerque, NM: University of New Mexico Press.

Voyles, Traci Brynne. 2015. *Wastelanding: Legacies of Uranium Mining in Navajo Country*. Minneapolis, MN: University of Minnesota Press.

Weber, Joe. 2021. 'The Lost Meridian of the 'Aha Makhav'. *Cartographica: The International Journal for Geographic Information and Geovisualization* 56(2): 106–19.

White, Paul J. 2006. 'Troubled Waters: Timbisha Shoshone, Miners, and Dispossession at Warm Spring', *IA: Journal of the Society for Industrial Archeology* 32(1): 4–24.

———. 2008. 'Chuckwalla and the Belligerent Burro: Timbisha Shoshone, Miners, and the Footprints of Dispossession in the Panamints'. Ph.D. dissertation, Brown University.
———. 2017. *The Archaeology of American Mining*. Gainesville, FL: University of Florida Press.
Williams, David. 1993. *The Georgia Gold Rush: Twenty-niners, Cherokees, and Gold Fever*. Columbia, SC: University of South Carolina Press.
Young, Otis E. 1970. *Western Mining*. Norman, OK: University of Oklahoma Press.

— Chapter 6 —

Contentious Legacies

Post-Mining and Heritage-Making in the Italian Alps

Roberta Clara Zanini

Introduction

The aim of this contribution is to present the outcomes of an ethnographic research conducted between 2011 and the present day in Macugnaga, a former mining community in the Anzasca Valley, in the Piedmontese Alps of north-western Italy. [1] Gold mining played a major role in the economy of the village from its inception in the mid-eighteenth century until the closure of the mine in 1961 (Zanini 2015). This event is still remembered as a major turning point that imposed a sudden conversion to a tourism economy, initially built on the alpine environment and subsequently, when the dominant model of winter tourism began to decline, on the culture.

The ethnographic data bring to light interesting politics and practices of heritage-making. In the following pages, we will see how these dynamics can be adequately understood if placed within the interpretative frame, now consolidated, that understands heritage-making processes as arenas within which the subjects involved activate often conflicting cultural, political and economic negotiations that contribute to redefining micro-local power relations (Del Mármol, Siniscalchi and Estrada 2016; Franquesa 2013; Meskell 2014). Macugnaga is both a (post-)mining and a linguistic minority community, being one of the Walser villages that occupy the high valleys at the foot of Monte Rosa. The coexistence of two available cultural assets has ignited competing instances of recognition and a *dual* process of heritage-making that intersects the even more complex issues linked to the uses of the alpine landscape as a resource for tourism. As far as the mining memory is concerned, this is leading to a process of heritage-making that looks *partial* in several acceptations of the term, as I will point out in the second section of this contribution, which outlines some theoretical reflections that have supported the ethnographic research and the subsequent anthropological analysis, and which will guide us in highlighting the process of heritage-making of Macugnaga's mining past. The next section will be devoted to the history of mining itself, since this path of rediscovery of memory is deeply rooted in the complex historical, economic and social dynamics linked to mining activity. The fourth part will explore the results of my ethnographic research to show how the process of recovering mining memory developed, in order to devote the fifth section to a more detailed examination of the frictions and conflicts that characterize the effort at patrimonialization we are focusing on. Finally, a brief conclusion will hint at a possible path towards recomposition.

This historical and ethnographical path will allow us to see how two different temporalities emerge: while seemingly separate, they constitute, on the contrary, two interrelated features of the same global minescape. For one thing, observing the temporal articulations of mining will make it possible to situate the local developments within global trajectories and to understand how demographic and migratory dynamics detectable at the micro level in a specific historically and geographically situated context are actually part of a broader series of connections and of a global historical process operating on multiple scalar levels (Clarence-Smith 2015). In addition, observing post-mining through the interpretative lens of heritage-making processes will allow us to broaden our gaze by including this second phase, only apparently disconnected from the previous one, in the same scenario, in order to verify whether, how and to what degree the dynamics observed in Macugnaga fit into what Michael Di Giovine (2009) calls the 'heritage-scape'.

Before delving into this complex and stratified process of heritage making, an excerpt from my field notes may prove useful in taking the reader on a quick stroll through the lanes of Pestarena, the hamlet of Macugnaga that has experienced the mining past with the greatest intensity:

> Everything in Pestarena speaks of the mine: at the entrance to the hamlet, there is a banner made by the association that works to preserve the memory of the mining past, and on which, in addition to the logo depicting the workers, the words 'Pestarena paese dei minatori' (Pestarena village of miners)

> stand out. In the main square, there is the usual emporium-bar, a tiny room that is intended to be the 'miner's house', with mining tools and equipment, and above all the monument to the miners who died in service, in memory of the dramatic accident on 13 February 1961 that caused the death of four miners and the subsequent closure of the mining plants. Observing Pestarena means seeing the traces of a past that is now distant: the administrative structures of the mine have only partially been restored, and as for the rest, they lie waiting for valid redevelopment projects that will make them accessible and usable by the community. A similar fate has befallen the entrance to the extraction wells, where winches and pulleys can still be seen, and the buildings that the administration made available to employees and their families during the period when mining made Pestarena the most populous hamlet and the economic hub around which the whole of Macugnaga and the Anzasca Valley revolved: infirmary, refectory, dance hall, cinema. Of all the facilities available, the only one that has retained its function is the municipal primary school. Macugnaga's pupils, from Pestarena to the upper hamlet of Pecetto, gather here in the building that in the past was attended by the numerous children of the miners. Its considerable size suggests the past number of children present in Pestarena was considerably higher than in the other hamlets – although these days a decreasing number of children come from Pestarena in favour of those from the upper hamlets. With its church dedicated to Saint John and the small cemetery full of tombstones with 'mining surnames', Pestarena now seems to be a faded copy of the populated and lively village that can be observed in the photographs kept in the 'miner's house'. (4 April 2011)

More than ten years have passed since my first descent into the field, to which this extract refers, but very little has changed. Macugnaga is no longer a mining community. The abandoned plants have only partially been restored, and no one has any serious inclination to resume an activity that represents the community's past and constitutes a definitively closed, although fundamental, economic chapter in the valley's history. Notwithstanding this, as we will see, Macugnaga is certainly a *living* 'community of mining memory' (Cappelletto 2003).

Partial Memories

James Clifford, in his now iconic *Introduction* to *Writing Cultures* (1986), has significantly contributed to highlighting how ethnographic writing is indeed a narrative production giving origin to what he called 'partial truths'. In particular, he noted that in the ethnographic process the poetic and political dimensions are inextricable. This consideration can be effectively applied to the process of heritage-making I deal with in this chapter. With this kind of heritage-making, poetics and politics need to be jointly and individually observed and analysed in order to understand their complexity and above all to detect their intrinsic 'partiality'.

My aim here is to highlight how these dynamics have developed through a long and still not complete process of negotiation and selection of meanings and tools, namely *what* to tell (which memory)? And *how* to communicate it? Through the protection and valorisation of *which* specific cultural assets? And, moreover, *who* is this narration addressed to? With regard to this last aspect, I have identified through the concepts of introversion and extroversion an analytical tool of considerable usefulness. Borrowing a formulation proposed by Jean-François Bayart (2000) and taken up by Alice Bellagamba (2009), it is possible to place on a continuum that separates extroversion and introversion the different politics and practices experienced within a community as far as the management of memory is concerned. It is thus possible to distinguish between introversive approaches, oriented towards the members of the community themselves and explicitly designed for them, and more extraversive approaches, aimed at communicating with the outside and with non-members of the group, and at exploiting and using memory as a resource for local economic development. I believe that both orientations are detectable in Macugnaga's memory management strategies. There are, in fact, simultaneously introversive and extraversive paths of memory; as we will see in this chapter, these two different modalities are not mutually exclusive. Rather, I think they can be considered as the two poles of a continuum along which the different cultural activities of the community are placed.

In Macugnaga the process of heritage-making of the mining community's past can be analysed by exploring the different meanings of the concept of partiality. Indeed, it is both a still incomplete process, and it is representative not of the whole local community, but only of a part of it. We will see in detail how the Walser people's different memory scenario is also non-neutral and rooted in partial representations of the different weight and importance of both Walser and mining components in Macugnaga's past. The adoption of this observational lens allows us to understand the numerous frictions that have characterized, in different ways, all the phases the community as a whole has experienced so far.

Macugnaga has an intense mining past and was populated by a German-speaking Walser minority. The head of the Anzasca Valley was settled as early as the thirteenth century by German-speaking peasants from Valais, as part of a wider movement of migration, colonization and reclamation of the highlands. This led to the presence of Walser settlements throughout the area on the southern and eastern slopes of Monte Rosa.[2] Macugnaga is therefore a linguistic minority community, and this characterization, in addition to being evident in both heteronomous representations and self-representations of the community itself, has been officially certified and protected since 1999, by virtue of the promulgation of Law no. 492/99 'Rules for the protection of historical linguistic minorities', which promotes actions to defend and enhance the Walser language, culture and memory. This is to say, Macugnaga is a community where this ancient German dialect is recognized, protected by the law and considered to be a fundamental cultural feature of the community. The law actually gave legitimacy to efforts to protect and safeguard the linguistic and cultural specificities that were not only widely prevalent at a local level – associations for the promotion of Walser culture were formed at the end of the 1970s – but also at a supranational level, thanks to the associative action of the Internationale Vereinigung für Walsertum (Führer 2002: 147–201).

Precisely because of the complex demographic transformations that have involved Macugnaga in the past, to which we will return later, different cores of meaning can now be identified in the local context, composing a complex mosaic of representations and memories that refer to the past and present of the community. These heterogeneous memories impose themselves in different ways and through the various – and always partial – paths of promotion and enhancement of the local cultural heritage on which I will focus.

I believe that what makes this process particularly significant is that in order to understand the dynamics of friction and conflict that emerge from ethnography, it is necessary to interpret them in light of the fact that heritage has become a hegemonic idiom. Both Jaume Franquesa, in a deep ethnographic account of the process of heritage-making in Majorca (2013), and Camila del Mármol, Valeria Siniscalchi and Ferran Estrada in a comparative study about the strategies of heritage-making in the Alps and in the Pyrenees (2016), have shown how 'heritage' gives structure not only to the narratives but also to the political and economic practices that are activated, irrespective of whether they adhere to hegemonic or, on the contrary, counter-hegemonic interpretations. During my fieldwork, I was able to detect in Macugnaga dynamics that were comparable to those highlighted in these two studies. It seems to me that the context I observed can now be seen, in its post-mining phase, as a frontier crossed by lines of friction and conflict (Tsing 2005). Those who may find themselves at a heritage-frontier are those upholders of a mining past who desire to obtain the legitimacy secured by being able to define the mining memory as 'heritage', in contrast, as we will see, to those who, like the members of the Walser community, already enjoy this and are placed, in the local heritage panorama, on a decidedly hegemonic level.

Before retracing the various phases of the process of mining heritage-making, which began only many years after the closure of the mine, it may be useful to devote the next section to the mining history of Macugnaga because the frictions detectable along the heritage-making path are deeply rooted in the social and cultural dynamics that characterized the community during the period of mining activity.

Two Centuries of Mining History

It goes without saying that any sound anthropological reflection on mining communities is bound to address issues of mobility, and this certainly applies to Macugnaga. Indeed, the presence of the largest Italian gold mine led to frequent and significant immigration movements throughout the whole activity period of Macugnaga's gold mine. Therefore, it is now necessary to highlight the ways in which these substantial migratory flows have historically manifested in the specific territorial area we are dealing with. In fact, although mining mobility obviously presents different characteristics and micro-dynamics depending on the specific contexts, it is nevertheless part of a common frame, which can be analysed and interpreted by utilizing a couple of concepts that proved to be strictly intertwined, namely knowledge and acquaintance. If we analyse the migratory movements that have profoundly modified community structures in mining contexts, the main vectors are professional skills (knowledge) and the networks of interpersonal relationships (acquaintances).[3] The dynamics observed in Macugnaga, therefore, are fully part of a broad reading of mining culture – and mining mobility – in Europe in general, and certainly in the Alps in particular.

As both the works of Riccardo Cerri and Alessandro Zanni (2008) and of Pier Paolo Viazzo (1989: 163–77; 2009) have pointed out, the development and spread of gold mining in Macugnaga in the second half of the eighteenth century was largely due to the planning and organizational action of mining entrepreneurs who had moved to the village at the head of the Anzasca Valley from nearby Alagna, another Walser community where proto-industrial mining was at a standstill at the time. This movement of the mining entrepreneurs around the slopes of Monte Rosa brought the migration of workers previously employed in Alagna, with the entrepreneurs having already had the opportunity to get to know them there. Historical investigations carried out on the archival documentation from the eighteenth century, although fragmentary, have shown that between 1760 and 1785 the workforce employed in the Anzasca mines numbered approximately 300, equal to about one third of the entire resident population of Macugnaga (Cerri and Zanni 2008). Almost all of these were foreign workers, who followed mobility trajectories based on the one hand on consolidated recruitment practices and on the other on the growth and decline of the mining industry experienced by different Alpine districts.

The example of Macugnaga confirms that it is not uncommon to find an overlap between miner and migrant. Moreover, historical research shows that the Anzasca Valley case is not isolated or connected only with the already mentioned Alagna but on the contrary extensive connections can be identified. In the past, in fact, mobilities of this kind involved various mining communities in the Piedmont and Aosta Valley district. An analysis of the documents contained in the municipal and parish archives, as well as the documentation produced by the mining companies, reveals numerous traces of the movements of workers from one mining town to another (Viazzo 1989: 163–77; 2009; Zanini 2015).

Investigating the origin and movements of the manpower employed in Macugnaga's mines requires a broader view, so as to include a sort of 'mining map'. If at the beginning of the mining activity skilled manpower came mainly from the Tyrol and from other areas of Piedmont, between the nineteenth and twentieth centuries the demographic increase was notable as a consequence of the arrival of miners coming from literally all over Italy, including Sardinia. It is interesting to note how the history of mining in Macugnaga is embedded in historical processes that transcend the strictly local and regional dimension, not only in terms of the labour force recruited but also in terms of ownership and capital investment. After a series of ups and downs, which gradually led to a considerable reduction in the workforce, during the second half of the nineteenth century the exploitation of the Monte Rosa goldfields became a fully-fledged part of European history. After a century of local entrepreneurship, British capital appeared on the Alpine scene, in the form of The

Pestarena United Gold Mining Company Ltd, founded in London in 1867, flanking colonial enterprises in South America with mining initiatives in the Alps. Mining capacity increased considerably following the large influx of foreign capital. Around 1880, the number of workers reached 350, but in the following decade it witnessed a rapid decline because of difficult extraction. The twentieth century began with a new change of ownership, with the mines returned to Italian hands after the British interlude. In 1939, at the height of the Fascist era and shortly before the outbreak of the Second World War, the Pestarena mines became part of the autarchic AMMI (Azienda Minerali Metallici Italiani), which intensified its recruitment of labour. Within five years, the number of miners employed rose from around 400 to almost 900 in 1943.[4]

It is important to distinguish between incoming and outgoing movements in Macugnaga's mining history. As we will see presently, during the twentieth century, Macugnaga also experienced a phase of massive 'mining emigration' due to the decline and then to the closure of the mines at the beginning of the 1960s, the cultural and symbolic repercussions of which we will focus on more closely. However, it should be emphasized that migratory movements in this locality, especially in the very early stages of its mining history, were much more frequently directed inwards than outwards. If we take a close look at the mobility trajectories that between the mid-eighteenth and mid-twentieth centuries linked Macugnaga to other mining communities in the Alpine arc, we can depict the image, which is not only metaphorical, of a space variously traversed by flows of manpower – that is to say, of a 'mining field'. The concept of a mining field was first used by David Levine and Keith Wrightson (1991) in what remains the most important historical, or at least historical-demographic, study of a mining community (Whickham, in the North of England). The two authors use this concept to define a large basin that transcended the territorial boundaries of a single village and whose socio-economic core was represented by the mining industry. In particular, what identified a mining field was mainly the fact that the workers employed in the mines were recruited not only in the village in which the mining plants were located but in a large surrounding area, the field, which therefore presented itself as a sort of reservoir of workers that was drawn on according to industrial needs. Depending on the scale level at which the observer stands, different mining fields of greater or lesser extension can be identified (Revel 1996). In the case of Macugnaga, the Anzasca Valley constituted what could be defined as the minimum level field, which in turn was included in gradually larger fields – such as the Piedmontese and Aosta Valley district – which were internally closely interconnected.

The early and intense dynamics of immigration radically affected Macugnaga's community, 'contaminating' its original Walser core. The analysis of archival documents highlights how in Macugnaga the immigrant miners and their families formed a 'subcommunity', separated by symbolic and spatial boundaries from the Walser community. To restrict ourselves to a spatial dimension, we can point out that the Walser population lived for the most part in the highest hamlets of the municipality – Staffa and Pecetto – while the miners and their families occupied the two lowest hamlets of Pestarena and Borca. In the upper part, the Walser population practised all activities that drew on the resources of the mountain environment: agricultural and pastoral activities, forestry and, above all, consolidated practices of hunting, smuggling and pack-hunting along the alpine routes and passes. In this period, therefore, the Walser component and the immigrant component of the population of Macugnaga seemed to merely coexist rather than actually live together.

The undulatory trend that characterized mining, and particularly gold mining, deeply shaped the composition of Macugnaga's population, with high rates of turnover that increased during the Fascist period, when mineral production went up again, stimulating the influx of miners not only from Alpine districts but from all over the country. These were the years when the mining administration built the facilities to accommodate the immigrant miners and their families in Pestarena, which profoundly changed the space and landscape of the village and put the hamlet at the core of the mining complex and of the valley as a whole.

Over the years, the increased presence of immigrant miners progressively broke the endogamic barrier and had serious repercussions on the linguistic profile of the community. If at first the cultural and symbolic boundaries between the Walser community and the mining community preserved the Walser linguistic identity, the massive immigration flows of the nineteenth and twentieth centuries radically changed the situation, leading to a considerable reduction in the number of Walser speakers within the total population. There is no doubt that the growing incidence of linguistically mixed marriages contributed to this decline (Zanini 2019).

Until 1961, when the mine closed, Macugnaga, and in particular its lower hamlet of Pestarena, was therefore undoubtedly a mining community. Macugnaga's mining history unfortunately came to a tragic end. Following a dramatic accident in a tunnel in which four miners lost their lives – significantly, they were not originally from Macugnaga but had all immigrated to the Anzasca Valley to work in the gold mines – the mining company took advantage of this tragedy to cease mining activity under the pretext of excessive danger, causing a large emigration of manpower. Actually, for some years, mining activity had been increasingly contracting. One reason is the exploitation that in previous decades had gone deeper and deeper entailed management and transport costs that were no longer sustainable. Second, at the end of the 1950s, an epoch of decline in mining activity had begun, involving not only the Ossola district but European mining in general, and the mines of the Alpine arc in particular. During the 1960s, this process intensified dramatically, and the closure of the Pestarena plant was part of this panorama: by the time it closed, the workforce had already been considerably reduced, and in 1960 it was limited to around 80 employees.

If we move from a local to a global level, this process of dismantling on a European continental level finds an interesting counterpart in the intensification of mining initiatives in sub-Saharan Africa, Asia and the Indian subcontinent. This movement of scale allows us to return to the concept of 'mining field' because it enables us to include in a single framework socio-demographic, historical and economic dynamics and processes that cannot be adequately analysed at a strictly local and regional level. On the contrary, considering the recruitment and mobility practices of mining workers at a global level makes it possible to grasp the intrinsic connections between processes of divestment and intensification in different areas at different times.

The emigration movements that characterized Macugnaga and the Anzasca Valley during the 1950s and 1960s can therefore be adequately included within this global framework. In fact, in the last phase of Macugnaga's mining history, as well as in the first years after the closure of the plants, many miners took the route of emigration, putting the skills they had acquired in the galleries under the Monte Rosa to good use in distant places such as the United Kingdom, Africa, India, Australia. Therefore, in its final stages Macugnaga's mining community became a recruitment pool of highly skilled and specialized manpower. In this phase too, movement away from Macugnaga constitutes an essential key to understanding the local sociocultural context. Beginning in the mid-twentieth century, both the geographies and temporal trajectories of mining in Macugnaga changed significantly. The temporal rupture caused by the closure of the mine fits seamlessly into a broader global temporal trajectory (D'Angelo and Pijpers 2018), which, when observed through a geographic lens, allows us to see how the global dynamics of mining exploitation result in dislocated workers, thus inextricably intertwining processes of dismantlement and capitalization both on the level of extraction and of heritagization.

The date of 13 February 1961, when the tragic accident occurred, marked a clear discontinuity of the socio-economic and demographic vicissitudes of the valley and a traumatic break in the history of the mining community, which from then on started falling apart. The ethnographic data show that the accident was a symbolic cut in the lives of the miners and their families. This abrupt interruption gradually turned into a virtual 'mining silence' that persisted until 2009, when the association 'Figli della Miniera – Children of the Mine' was founded. We can identify two factors that led to the establishment of the association. First, the push constituted by the publication of an imposing study on Macugnaga's mining history (Cerri and Zanni 2008) was decisive in giving scientific legitimacy to the local mining past. No less important was the endogenous boost, from the commitment of an elderly ex-miner, Angelo Iacchini, in amassing a collection of mining tools and testimonies, which had begun with the closure of the mine.

Two Decades of Mining Memory

After the foundation of the association, therefore, and notwithstanding the gradual erosion of the mining community, a new sort of community came to light, a 'community of memory', which is maintained, renewed and regenerated thanks to the commitment of its members to passing on cultural heritage relating to the mining past and reuniting on commemorative occasions. Commemorating the 1961 traumatic event and mourning are fundamental to this 'mnemonic community', with the same sort of ceremony attached to the date as remembrance of Second World War atrocities (Cappelletto 2003, 2006; Dei 2007).

Thus, the heritage-making process observed in Macugnaga began with the foundation of the association and developed around a dialectic between introverted and extroverted memory (Bayart 2000; Bellagamba 2009). Observing this dialectic allows us to shed light on the frictions I referred to at the beginning of this chapter. In the very first phase, the activities of 'The Children of the Mine' had no tourist purpose, and were mainly introverted, on two different levels, both with regard to the mining memory community specifically and the Macugnaga community as a whole. The aim was the 'maieutic' recovery of memory, in order to recollect contents and meanings that for a very long time had remained in the background within the mining community itself. The 'mining silence', which began after the closure of the mine, had in fact intensified since the late 1970s. At that time the revival of linguistic minorities was gaining ground (Boissevain 1992; Führer 2002), helping to delineate a frame within which the Walser component of the local community could find cultural legitimacy, eventually initiating a process of identity consolidation that in the following years gradually led to the identification of the Walser tradition as Macugnaga's main historical and cultural characterization, with a tendency to be exclusivist or at least hegemonic.

In this first introverted phase, frictions emerged mainly on the level of symbolic representation and were connected in the first instance to the recognition of the historical role played by the mine and its workers in giving shape, resources and life to Macugnaga. Ethnographic research revealed the clash between the positions of those who see this past as economically and socially all-important and those who do not – that is, part of the population of Walser origin, which instead interprets mining as a sort of 'historical accident' that contaminated the Walser culture, still considered by far the more important and foundational culture.

During my fieldwork, I had the opportunity to directly observe the activities of the association and to participate in the commemorative events organized by its members. A very complex cultural and symbolic panorama emerges that finds in the representations of the figure of the miner and in the decisive role of the migratory experience crucial themes for understanding, on the one hand, the dynamics of transmission and valorisation of memory and, on the other, the sociocultural dynamics within the community itself.

If the intent is to propose an analysis of the representations of the miner emerging from the process of memory recovery, it is first necessary to clearly distinguish between self-representations – an expression of the miners themselves or, at a slightly more general level, of the mining community as a whole – and heteronomous representations (Zanini 2016). The heteronomous representations, sometimes conveyed by members of that part of the Macugnaga community not directly involved in mining or even by those outside the community itself, not infrequently tend to reproduce a miserabilist interpretation of the figure of the miner and of mining work. From this point of view, the migratory experience of leaving the community after the mine's closure to look for work abroad is read as a defeat, a forced choice, the inevitable conclusion of an individual history interpreted more or less implicitly by resorting to a deterministic-constrictive representation. Just as these same representations explain the decision to become a miner as a 'non-choice' imposed by contingency and economic necessity due to the constraints of the local context, similarly they interpret the miner's becoming a migrant as the obvious conclusion of a destiny marked by poverty (Zanini and Viazzo 2015). One of my most outspoken witnesses involved in local heritage policies, a woman of Walser origin who is very committed to the promotion of Walser culture, expressed herself on this subject in caustic and almost contemptuous terms: 'I honestly wonder what there is to remember about that past. I have them in mind, these poor men who came out of the mine dirty, destroyed. I think we should forget, rather.'

This testimony is particularly revealing because it brings the issue of the minescape's different temporalities back to centre stage, highlighting how even in the transition from mining to post-mining the representations of culturally hegemonic groups, of which our witness is evidently a member, continue to imagine miners as culturally subaltern historical remnants, illegitimate new entries in the heritage economy arena. Completely different – opposite, I would say – are what we can characterize, at various levels, as mining self-representations. One of the most interesting features of the 'Children of the Mine' is the fact that it brings together three generations of memory bearers, namely the very few elderly miners still alive, their children and their grandchildren. These three generations, who have experienced life in the mines in radically different ways (or, in the case of the grandchildren, not at all), consequently express different contents, which find hospitality in the numerous commemorative events promoted by the association and always attended in a very participatory way by its members. Two recurring themes clearly emerge, namely the risk associated with mining and the experience of migration. As I noticed, however, the modulations with which these two themes are presented may differ quite considerably depending on the generation to which the interlocutor belongs.

The children's generation, which is numerically the most significant and constitutes the heart of the association, is much more focused on the theme of risk, pain and suffering that mining life imposed on miners and their families. The grandchildren are slightly different, being more oriented towards the organizational activities of the association and the enhancement of the mining history of the community. Still different are those who actually worked in the mine (who, it should be emphasized, present the only 'genuine' *self*-representations), who certainly make constant reference to the theme of risk but contextualize it in representations that convey the image of the miner as a capable professional. In this sense, the theme of risk is inextricably linked, on the one hand, to that of migration and, on the other, to that of the professional competence needed to take care of oneself, one's family and one's fellow workers.

The ethnographic evidence highlights how migration constitutes a fundamental stage in what we might call the anthropopoietic (Remotti 2009: 277–341) construction of the miner and how the representations of the figure of the miner himself and of mining work in general are closely linked to this theme. My ethnographic research has revealed the importance of the migratory dimension in itself, as well as the economic, but also symbolic evaluations that oriented the miners' choices, such as the observation that working in the tunnels guaranteed a standard of living considered adequate for one's family, or the decision to emigrate. Therefore, it seems legitimate to contend that what has been defined by many as a 'mining culture' (Sanga and Viazzo 2016) presents features very similar to those outlined by the now well-established theoretical reflections on the 'culture of migration' (Cohen 2004), starting from the complex interweaving of cultural constraint and individual choice in defining professional and migratory strategies and projects.

One of the most frequent mining narratives in Macugnaga was that of the experience of emigration to Africa. Between the late 1950s and the early 1960s, when the decline of the Macugnaga mine became apparent not only to the management but also to those working in the gallery, a large number of miners were recruited to go to Africa to work in the mines in the area between Uganda and Tanzania; the memory of this recruitment remains very vivid for those involved:

> There was a great demand. Through the Macugnaga Chamber of Labour, about twenty of us, all from Macugnaga, went down to Rome to make the selections. There was a lot of coming and going: some went to Africa, others to Belgium, others to America. They took us on because we were all good people for the job, whether in the factory or in the gallery, we were all already selected for the job.

The decision to emigrate, even though it was possible to turn to alternative local activities, such as, for example, the emerging construction of ski facilities, was closely linked to the economic potential of the migratory experience but was not separated from more strictly cultural and symbolic factors. Working abroad, especially in Africa, ensured a sufficiently high income to make emigration not a second-best solution but on the contrary a conscious and carefully considered choice. In particular, it guaranteed the kind of economic capital that, once returned, could be invested in the form of local real estate. This economic capital therefore meant considerable symbolic capital. Being able to buy a house and ensure a decent life for one's family was proof of the success of emigration and of one's value as a miner and as a man, as this testimony eloquently explains:

> Oh, Africa was good! Ah, everyone who went to Africa came home and made a little house or bought a house. I used to send home 50 pounds a month through the bank. The bank in Nairobi sent it to the bank in Italy and my mother went to pick up the money and put it aside for me in the bank, she didn't need it and I had given her the proxy to go and get it. She didn't need it, and when I got home, I had a little bit of money. I bought a house here, I built a house in Quarazza [another small hamlet in the Macugnaga area], in short I sorted things out.

After almost twenty years of painstaking maieutic recovery of the mining memory, and ten years after the foundation of the association, although this multiplicity of representations is not completely recomposed into a single, fully shared narrative, the 'community of mining memory' has found a rather defined and accepted space within the local sociocultural context. Consequently, in the last period, the activities of the association have started to mitigate the introverted approach of the early times in favour of a more extroverted orientation, aimed at promotion and valorisation outside the local community.

A Still Incomplete Process of Heritage-Making

In this regard, the frictions highlighted before intertwined with the complex dynamics of local politics. In fact, the competition shifted from the symbolic level of the recognition of the role of the mine in the past, to the much more concrete division of the economic resources for the enhancement of the local cultural assets in the tourism sector. Between the 1960s and the beginning of the 1980s, the local area was geared mainly towards winter tourism, built on the exceptional landscape and environmental resources guaranteed by the presence of the majestic East Face of Monte Rosa.

In order to understand the assumptions on which Macugnaga's tourist development was based, it is necessary to metaphorically leave the mining tunnels of Pestarena and look up at the Monte Rosa massif; the presence of the dominant East Face has had a decisive influence on Macugnaga's history in terms of tourism. This influence, however, did not only manifest itself at the time when the decline of mining imposed an economic reconversion but had its roots in previous centuries. Before Macugnaga became famous as a ski resort in the 1960s, it had already accumulated almost two centuries of mountaineering history and culture (Hansen 2013). The practice of mountaineering had found fertile ground at the foot of Monte Rosa precisely because of the environmental, social and cultural context of Macugnaga. So, it was in a context in which the practice of mountaineering and the related professions – hotelier, mountain guide – were strongly rooted that mass tourism took hold during the second half of the twentieth century. The number of hotels multiplied, residential construction grew, the number of local mountain guides increased considerably, and ski schools were set up. Between the end of the 1950s and the late 1960s, while the mine was running out of capacity to economically support the village, the ski facilities were inaugurated and a new economic and social phase began for the community, which would mark the increasing contraction of traditional agricultural activities and the expansion of tourism, even if the resources to be used for promotion and enhancement would change over time.

In fact, when winter tourism began its downward trend, during the 1980s, the community looked at 'Walser ethnicity' as an available resource for cultural tourism, with conflict patterns and dynamics very similar to those highlighted by Davydd Greenwood in his pioneering study of the Fuenterrabia community in the Basque Country (1977). These local dynamics of patrimonialization are embedded in a broader theoretical framework that in the last decade of the twentieth century profoundly characterized not only the public discourse and policies of cultural intervention but more generally Alpine anthropology as a disciplinary field. Interest moved first towards 'tangible cultural heritage' – that is, buildings, objects or sites with historical and cultural value – and then – especially following the approval of the UNESCO 2003 Convention for the Safeguarding of Intangible Cultural Heritage[5] – towards some of the broad domains where intangible cultural heritage was likely to manifest, such as rituals and festive events, traditional craftsmanship, and knowledge and practices concerning nature (Arizpe and Amescua 2013; Meskell 2014). This new conversion led to the economic capitalization of the Walser cultural heritage consistent with a representation of the mountain as a space of authenticity and conservation of traditional cultures. Further evidence of how the process of enhancing the value of Walser culture is being pursued through the hegemonic idiom of heritage is provided by the fact that the Walser Communities in the Alpine arc have recently embarked on the path of applying for the inclusion of Walser culture within the UNESCO list.

It does not therefore seem unfounded to state that the Walser part of the community, in the last thirty years, has substantially monopolized the available resources and spaces for promotion and enhancement, to the detriment of the 'younger' mining memory community, which still struggles to find spaces, ways and resources to further its cultural heritage. Indeed, the struggle of the mining memory community is precisely aimed at finally getting into and settling *within* the hegemonic scenario of heritage, to acquire the legitimacy that heritage labelling guarantees (Franquesa 2013) – that is to say, to move from the level of mere local symbolic recognition to the decidedly far more powerful and protected – politically and economically – level of heritage-scape (Di Giovine 2009).

The difficulties that the mining memory community is encountering along the process of patrimonialization, both in terms of intangible memory and tangible heritage, are largely due to micro-political relations linking – or, better, *not* linking – the 'Children of the Mine' association with local administrations. Over the years, in fact, the association has received what we might describe as formal 'rhetorical' support from local administrations and institutions, but this has never translated into real operational support, in terms of heritage-making and economic investment policies in the cultural tourism sector or in terms of future planning. While it is not possible to identify activities or projects explicitly oriented towards boycotting the consolidation of a mining patrimonialization (which, even on a symbolic level, would be controversial and locally unacceptable), we cannot point to any real investment in its support. It is precisely this 'abandonment' by the institutions, and the feeling of loneliness that derives from it, that the members of the association lament with particular bitterness, as it makes their heritagization efforts look very demanding, solitary and almost amateurish. In truth, it should be noted that the process of heritage-making of the mining legacy has actually followed two different paths. Although not without difficulties the process of valorisation of the *intangible* heritage connected to the mining past has in fact gradually strengthened, the course is still decidedly partial, and limited, as regards the *material* and *architectural* heritage.

With regard to intangible legacy, I was able to observe a growing, albeit laborious (and always rather marginal in the overall community scenario), activity of collection, enhancement and transmission of memories, narratives and knowledge related to mining that makes itself visible mainly during events linked to the local festive calendar: celebration of the village patron saint, summer fairs, Saint Barbara's Day (the feast of the miners' patroness). Furthermore, the association's most recent activities have been explicitly aimed at reconnecting Macugnaga with other Italian and European mining communities, in particular to re-establish the relationship between the hamlet of Pestarena and the mining communities that were tied to it both as recruitment basins and as destinations of the miners from Macugnaga. The institutional and cultural twinning among these areas gives origin to a symbolic movement from a local memory community to a wider, even supranational, memory community that retraces the actual movements of the manpower that served in Macugnaga's gold mine during the twentieth century.

As I pointed out, the intertwining of the two approaches, introversion and extroversion, is at the heart of the association's activities and finds in the issue of migration a fundamental thematic core, both in dissemination and promotion projects and in research projects. The memory of migration, the shared narration of one's own migration experiences and of those of one's fathers or grandparents, is actually a narrative canvas that can be found, together with the similarly pervasive considerations on the theme of risk (Zanini and Viazzo 2015), at almost all public events related to the association's work. At the same time, historical research activities, which originally constituted the stimulus that led to the foundation of the association itself (Cerri and Zanni 2008), and which still constitute the scientific backbone of the association's work, continue to identify mobility as an essential thematic core. While in the first phase, however, the focus was on the migratory movements that modified the community structure during the eighteenth century, the most recent investigations have focused on the still little explored emigrations at the beginning of the twentieth century, with particular attention to the flow of miners who moved from Pestarena to the United Kingdom to work in the mines of Frongoch, in Wales, and Nenthead, in Cumbria, where the Société des Mines et Fonderie de Zinc de la Vieille Montagne, the Belgian multinational that owned the most productive lead and zinc mines in Europe in the late nineteenth and early twentieth centuries, was active. [6] Once again, therefore, we see the emergence of a 'mining field', which in this case we could perhaps more appropriately define as a 'mining *memory* field', which shows us how the theme of mining is always intertwined with that of migration, of symbolic connections and, more generally, of movement – whether of professional knowledge or of a memory to be transmitted and enhanced.

As far as the material and architectural heritage is concerned, the process has so far been limited to the recovery of a small sector of the tunnels and of the administration buildings but without any plans for revitalization, neither in a museum perspective nor in other ways. It is currently possible to visit only one tunnel, privately owned, which hosts a small historical-technological exhibition that unfortunately falls far short of the standards reached by comparable exhibitions in other European mining districts. Furthermore, this recovery has not been integrated into a coherent planning effort concerning the whole mining complex, or with activities of valorisation of the intangible heritage carried out by the association 'Children of the Mine – Figli della Miniera'.

The most evident element that emerges clearly from both the ethnographic research and from comparisons of the heritage-making process activated in Macugnaga with those that one can observe in other mining districts both nationally and internationally is the lack of a medium to long-term project organized at the local level in an integrated, coherent, scientifically grounded and institutionally supported scenario. On the contrary, the numerous minute events designed to promote mining memory do not manage to go beyond the strictly local dimension to find a place within a global minescape, precisely because of the lack of support from institutions, not only economically but especially politically, in the deepest sense of the term. The historical and archival research activities, although of undisputable scientific value, are conducted on a solitary, voluntary basis and do not find adequate space and recognition in local planning. Furthermore, the twinning agreements I referred to, which seem to envisage the possibility of building a transnational collective memory, have not yet been converted into sound planning practices that could effectively place the memory within a global scenario of valorisation of the mining past.

Conclusion

The reasons for this unaccomplished process of revitalization can be detected at two different levels. On the one hand, a major obstacle is patently represented by the lack of economic resources that are predictably huge and well beyond a voluntary association's reach. On the other hand, but no less important, the architectural recovery of the mine does not fit with the representation that the majority of Macugnaga's inhabitants mean to convey about its natural and anthropic landscape. There is no real desire to invest, economically but also symbolically, in the enhancement of elements that do not adhere to the aesthetics of the alpine landscape closely linked to its natural resources and to the traditional Walser forms of architecture. Hence, we can detect another friction between the process of mining heritage-making and the maintenance of a model of alpine tourism still very rooted in the winter exploitation of environmental resources and in the cultural traditions of the linguistic minority group, which represents itself, and is often represented from outside, as a numerical majority.

Notwithstanding this seemingly irreducible friction, my last ethnographic observations allow me to pinpoint an experiment that appears to suggest the possibility of a recomposition. In fact, in November 2019, the 'Children of the Mine' association celebrated its tenth birthday by inaugurating 'The Miners' Trail', a trekking path with explanatory panels that unravel through a series of historically significant points of the upper Anzasca Valley mining site. The explicit purpose of this effort is to try a new way to integrate the mining heritage into the local tourist offer. As the crisis of winter tourism looks apparently irreversible, it is essential to find new forms of sustainable exploitation of environmental resources and to promote forms of soft, low-impact tourism. The newborn 'Miners' Trail' seems to be going in this direction, by integrating the transmission and enhancement of the cultural mining heritage within the valorisation of the Alpine landscape and of the mountain environment, which seems to constitute the only, truly common community horizon. Still, it is an initiative that retains a predominantly local dimension. In conclusion, while the Walser community, also by virtue of an explicit adherence to the UNESCO frame, has managed to find its own positioning within the heritage-scape, the same cannot be said for the mining memory community. The subaltern positioning persists despite the struggle of only seemingly paradoxical attempts to bring a counter-hegemonic narrative into the local heritage discourse precisely by adhering to the hegemonic idiom of heritage. The truly global life of Macugnaga's mining history does not seem to have yet been translated into a global post-mining memory.

Roberta Clara Zanini holds a Ph.D. in Anthropological Sciences from the University of Turin and is currently Research Fellow at the Department of Foreign Languages, Literatures and Modern Cultures at the same university. She has taken part in several projects funded by the University of Turin and the EU concerning the demographic changes involving the Alpine area and their social and cultural effects. She has long experience in ethnographic and anthropological research with a specialization in Alpine anthropology, anthropology of linguistic minority communities and mining communities and anthropology of welfare services in marginal areas.

Notes

1. This piece draws on material collected over a period of more than ten years of ethnographic investigation, initially conducted as part of my doctoral research in Anthropological Sciences at the University of Turin, in the frame of the Interreg Project E.CH.I. (Italo-Swiss Ethnographies for the valorisation of the intangible heritage) and subsequently of the LIMINAL Project – Linguistic Minorities in the Alps: Ethnicity, Languages and Demographic Processes (University of Turin).
2. The literature on the settlements founded in the thirteenth century onwards by colonists from Valais in many valleys of the Alpine arc, especially in Piedmont and the Aosta Valley and in Graubünden, but also in Liechtenstein, Vorarlberg and Tyrol, is very extensive. A general outline of the history of Walser colonization in Italy is offered by the seminal work of Paul Zinsli (1968). In-depth studies of the historical path that led to the presence of German-speaking groups in the area around Macugnaga are those of Mortarotti (1979) and Zanzi and Rizzi (1988). On the specific case of Macugnaga, see the volume by Zanzi, Rizzi and Valsesia (2006).
3. Similar reflections have also been proposed in historical studies with a broader comparative intent, such as the work by Donata Brianta (2007) on the movement of mining elites and the transfer of technologies in modern-day Europe.
4. I am deeply grateful to Riccardo Cerri, who generously gave me the opportunity to refer to his most recent historical research, which although publicly presented at events organized by the association 'Children of the Mine' and through the official organ of the association itself ('Il Foglio d'oro', years 2017 and 2019) is still largely unpublished.
5. The UNESCO Convention, signed in Paris on 17 October 2003, is available at http://portal.unesco.org/en/ev.php-URL_ID=17716&URL_DO=DO_TOPIC&URL_SECTION=201.html.
6. See above, Note 3.

References

Arizpe, Lourdes, and Cristina Amescua (eds). 2013. *Anthropological Perspectives on Intangible Cultural Heritage*. Dordrecht: Springer.

Bayart, Jean-François. 2000. 'Africa in the World: A History of Extraversion', *African Affairs* 99(395): 217–67.

Bellagamba, Alice. 2009. 'After Abolition: Metaphors of Slavery in the Political History of the Gambia', in Benedetta Rossi (ed.), *Reconfiguring Slavery: West African Trajectories*. Liverpool: Liverpool University Press, pp. 63–84.
Boissevain, Jeremy (ed.). 1992. *Revitalizing European Rituals*. London: Routledge.
Brianta, Donata. 2007. *Europa mineraria: Circolazione delle élites e trasferimento tecnologico (Secolo XVIII–XIX)*. Milano: FrancoAngeli.
Cappelletto, Francesca. 2003. 'Long-Term Memory of Extreme Events: From Autobiography to History', *The Journal of the Royal Anthropological Institute* 9: 241–60.
______. 2006. 'Social Relations and War Remembrance: Second World War Atrocities in Rural Tuscan Villages', *History and Anthropology* 17(3): 245–66.
Cerri, Riccardo, and Alessandro Zanni. 2008. *L'oro del Rosa: Le miniere aurifere tra Ossola e Valsesia nel Settecento*. Magenta: Zeisciu Centro Studi.
Clifford, James. 1986. 'Introduction: Partial Truths', in James Clifford and George E. Marcus (eds), *Writing Culture: The Poetics and Politics of Ethnography*. Berkeley, CA: University of California Press, pp. 1–26.
Cohen, Jeffrey H. 2004. *The Culture of Migration in Southern Mexico*. Austin, TX: University of Texas Press.
Clarence-Smith, W.G. (ed.). 2015. 'Mining in Global History', special issue of *Journal of Global History* 10(1): 1–76.
D'Angelo, Lorenzo, and Robert J. Pijpers (eds). 2018. 'Special Section: Mining Temporalities: Extractive Industries and the Politics of Time', *The Extractive Industries and Society* 5(2).
Dei, Fabio. 2007. 'Storia, memoria e ricerca antropologica', in Clara Gallini and Gino Satta (eds), *Incontri etnografici: Processi cognitivi e relazionali nella ricerca sul campo*. Roma: Meltemi, pp. 40–67.
Del Mármol, Camila, Valeria Siniscalchi and Ferran Estrada. 2016. 'Reflecting on Heritage and Power: Dynamics, Strategies and Appropriations in the Catalan Pyrenees and the French Alps', *International Journal of Heritage Studies* 22(5): 341–54.
Di Giovine, Michael. 2009. *The Heritage-scape: UNESCO, World Heritage and Tourism*. Lanham, MD: Lexington Books.
Franquesa, Jaume. 2013. 'On Keeping and Selling: The Political Economy of Heritage Making in Contemporary Spain', *Current Anthropology* 54(3): 346–69.
Führer, Johannes. 2002. *Die Südwalser im 20. Jahrhundert*. Brig: Geschichtsforschender Verein Oberwallis.
Greenwood, Davydd. 1977. 'Culture by the Pound: An Anthropological Perspective on Tourism as Cultural Commoditization', in Valene L. Smith (ed.), *Hosts and Guests: The Anthropology of Tourism*. Philadelphia: University of Pennsylvania Press, pp. 129–38.
Hansen, Peter H. 2013. *The Summits of Modern Man: Mountaineering after the Enlightenment*. Cambridge, MA: Harvard University Press.
Levine, David, and Keith Wrightson. 1991. *The Making of an Industrial Society: Whickham, 1560–1765*. Oxford: Clarendon Press.
Meskell, Lynn. 2014. 'States of Conservation: Protection, Politics, and Pacting within UNESCO's World Heritage Committee', *Anthropological Quarterly* 87(1): 217–44.
Mortarotti, Renzo. 1979. *I Walser nella Val d'Ossola: Le colonie tedesco-vallesane di Macugnaga, Formazza, Agaro, Salecchio, Ornavasso e Migiandone*. Domodossola: Giovannacci.
Remotti, Francesco. 2009. *Noi, primitivi: Lo specchio dell'antropologia*, 2nd edn. Torino: Bollati Boringhieri.
Revel, Jacques (ed.). 1996. *Jeux d'échelles: La micro-analyse à l'expérience*. Paris: Gallimard-Le Seuil.
Sanga, Glauco, and Pier Paolo Viazzo (eds). 2016. *La cultura dei minatori delle Alpi* [= *La Ricerca Folklorica* 71]. Brescia: Grafo.
Tsing, Anna. 2005. *Frictions: An Ethnography of Global Connection*. Princeton, NJ: Princeton University Press.

Viazzo, Pier Paolo. 1989. *Upland Communities: Environment, Population and Social Structure in the Alps since the Sixteenth Century*. Cambridge: Cambridge University Press.

———. 2009. 'Le comunità walser del Monte Rosa tra XVIII e XIX secolo: demografia, economia e migrazioni', in Pier Paolo Viazzo and Riccardo Cerri (eds), *Da montagna a montagna: Mobilità e migrazioni interne nelle Alpi italiane (secoli XVII–XIX)*. Alagna Valsesia/Magenta: Zeisciu Centro Studi, pp. 65–83.

Zanini, Roberta Clara. 2015. *Salutami il sasso: Dinamiche della popolazione e della memoria in una comunità alpina di confine*. Milano: FrancoAngeli.

———. 2016. 'Perché minatore? Il lavoro di miniera tra stereotipi, memorie e rappresentazioni', *La Ricerca Folklorica* 71: 81–98.

———. 2019. 'Migrazioni minerarie nelle comunità walser piemontesi: percorsi, effetti, rappresentazioni', in Michela Del Savio, Aline Pons and Matteo Rivoira (eds), *Lingue e migranti nell'area alpina e subalpina occidentale*. Alessandria: Edizioni dell'Orso, pp. 75–92.

Zanini, Roberta Clara, and Pier Paolo Viazzo. 2015. 'Il lavoro di miniera nelle Alpi: un'antica attività industriale tra realtà e rappresentazioni antagonistiche', *Popolazione e Storia* 15(1): 117–31.

Zanzi, Luigi, and Enrico Rizzi. 1988. *I Walser nella storia delle Alpi: Un modello di civilizzazione e i suoi problemi metodologici*. Milano: Jaca Books.

Zanzi, Luigi, Enrico Rizzi and Teresio Valsesia. 2006. *Storia di Macugnaga*. Domodossola: Fondazione Maria Giussani Bernasconi and Fondazione Enrico Monti.

Zinsli, Paul. 1968. *Walser Volkstum in der Schweiz, in Vorarlberg, Liechtenstein und Piemont*. Frauenfeld-Stuttgart: Huber.

— Chapter 7 —

Uranium Mining in New Mexico

Global Entanglements, Earth Relations and Awkward Ways

Targol Mesbah

'Everything for Everyone and Nothing for Ourselves'

'Everything for everyone and nothing for ourselves', she said slowly, letting the fullness of each word resonate in the large auditorium. She repeated: 'Everything for everyone and nothing for ourselves.' It was the second day of the International Uranium Film Festival at the Navajo Nation Museum in autumn of 2018.[1] Radmilla Cody opened the morning session with her words and song. She introduced herself as being from the Red Bottom People clan and an African-American woman. Cody performed the song 'Four Sacred Elements' and welcomed us, on this unexpectedly snow-covered morning, to Diné territory. She then recounted a little-known history: on another morning in July 1979, an earthen dam, filled with waste from the milling of uranium ore, gave way and released 1,100 tons of radioactive sand and 95 million gallons of contaminated mill liquid into the Rio Puerco river system in Church Rock, a town about 48 kilometres southwest of where we were attending the film festival.

There were other stories of contaminated waterways, food ways and life ways shared from other geographies during the three-day film festival. Similar but different stories. Stories shared through films, open discussions, personal conversations and over meals. We were an audience of international and local filmmakers, artists, researchers, activists and community members affected by the imposition of unearthed uranium from mining, milling, transporting, processing, waste or weapons. There may have been others in attendance. But Cody spoke to all of us. 'Extractive capitalism continues to exist because it provides the false hope of economic security through the breaking up of kinship with each other and the land', and 'it is in fact the collective will and kinship of the good people of the earth that will create change'. As she was coming to a close, Cody paused to reflect on how this notion of the collective will and kinship, K'é in Diné, is not just a word but imparts lived values and teachings she was raised under by her grandmother.[2]

This chapter locates the legacy of uranium mining in New Mexico as part of a global nuclear fuel chain that disproportionately affects the lives of indigenous and poor populations. Nuclear colonialism names a system of domination sustained through the nuclear chain of production, from uranium mining to nuclear fuel and weapons manufacturing, and reflects how those most affected are excluded from decisions made by nuclear states and corporations, which subject them to harm (Churchill and LaDuke 1986; Kuletz 1998).[3] In fact, the nuclear state is necessarily non-democratic because of the secrecy around the science, technology and uranium material needed for the production of nuclear fuel and bombs (Kohso 2020; Scarry 2014). Even as nuclear colonialism *disproportionately* affects populations who are historically excluded from participating in the circulation of power, the governance structure of the non-democratic nuclear state affects everyone – whether from the threat of nuclear war or the slow violence (Nixon 2011) of radiation exposure inherent in the process of mining uranium and all the subsequent stages of its handling such as milling, transport and waste storage. Radionuclides – radioactive forms of elements – do not abide by imposed borders; they move through the web of life, waterways, soil and food ways, with the winds, and cause genetic mutations in our children and their children, across generations.[4]

In this book, the 'global life of mines' offers a framework for the multiple temporal scales that connect mining and post-mining across different spaces. Pusceddu and Zerilli invite us to take up the global life of mines for a methodology that can engage the many and intimate interconnections between resource extraction and its afterlives. Here, I focus on the legacies of uranium mining in New Mexico to study its layered connections to a global chain of nuclear production and its multiple connections to the radioactive afterlives of uranium in the fabrication of nuclear energy, weapons and their waste. I am also curious about what futures can overcome the toxic legacies and afterlives of radioactive contamination. For this reason –

for a dignified future – my approach to the entanglement of global capital and colonialism follows the Zapatista mobilization of global resistance from below, which I will describe momentarily. I consider what ways of thinking, writing and acting in the world are helpful at this moment of intensified racist authoritarian (nuclear) states and climate crisis. How do I/we resist reproducing extractivist relations of power in pursuing research? This is the question that informs my research methods, which sometimes feel awkward and in some cases seek the awkward in order to shake off sedimented ideas around research. How to loosen the grip of hardened binaries of indigenous/Western or culture/nature in my own thinking, to see new connections and see anew existing ones, even if at times unsettling? Silvia Rivera Cusicanqui proposes we pay attention to the political economy of knowledge production and describes how North American universities 'have adopted the idea of subaltern studies and launched debates in Latin America, thus creating jargon, a conceptual apparatus, and forms of reference and counter reference that have isolated academics themselves from any obligation to dialogue with insurgent social forces' (Cusicanqui 2020: 51).

Cusicanqui writes about the practices and discourses of decolonization within the context of Andean histories of resistance and indigenous modernity. More specifically, she critiques how North American universities canonize and reproduce hierarchies of texts and authors within a system that remains ultimately separated from the social upheavals and insurgent movements aimed at changing the material distribution of the conditions of life and liveability. This critique informs my research and writing as I turn to critical theories, histories and ethnographies, participant observation, journalism, epidemiological studies, conversations with activists, scientists and others, The International Uranium Film Festival, specific films, direct actions, zines, teach-ins, webinars, peer-reviewed academic publications, self-published texts, United States Department of Defense publications, the political analysis of the Zapatista Army of National Liberation, and the teachings of grandmothers.

I came to research uranium mining in the Southwest of the United States by way of the Middle East. My family is from Shiraz in southern Iran; I was born and raised in Tehran, the capital, and immigrated to the United States with my family during the Iran-Iraq War of the 1980s. Like Cody and many others, I was raised by my grandmother's teachings but transmitted through my father, as she passed away before I was born. She raised her children with the words and wisdom of the thirteenth-century poet Sa'di of Shiraz. His verses are some of the most familiar in the Persian language oral tradition and describe an orientation to the world, a way of relating, to self and others. Perhaps the most well-known verse is from his Rose Garden collection of stories and poems describing what it means to be human.

> Human beings belong to each other's limbs
> For they are created from one precious essence
> If a limb falls to pain one day
> The other limbs remain in disarray
> You who are not touched by the others' pain
> A Human you cannot be named

In this poem, Sa'di of Shiraz is advising an unnamed King on matters of governance, and even if the succession of dynasties and dictatorships before and since the poet's time have failed to reflect this orientation in their actions, these words have not ceased to pass across the lips and hearts of the people. Oftentimes, reciting the first line is enough to invoke the desired quality of feeling oneself in relation to another in an interconnected web of life. This orientation informs how I come to the subject of my research on uranium mining and how we might collectively overcome nuclear colonialism.

'Everything for everyone and nothing for ourselves' is one of the Zapatista expressions for a dignified life that envisions a world they have been struggling for and building since before they recuperated territories in January 1994 in Mexico. The Zapatista Mayan peasant movement in the south-east state of Chiapas has been narrating their resistance to and analysis of 500 years of capitalism and colonialism since that time (Sixth Commission 2016). They share their practices of autonomy through online communiques, speeches, declarations, seminars and international encounters and continue to build their autonomous communities guided by the seven principles of *Mandar Obedeciendo* (to govern by obeying), a practice of governance that follows the will and participation of the people. 'Everything for everyone and nothing for ourselves' echoes the ethos of these principles of horizontality, reciprocity and dignity. Cody's opening remarks captured a similar vision of a world that struggles for 'a dignified and true life' for everyone. Shared too is an analysis of how the most basic conditions of liveability are threatened by capitalism and colonialism as systems that harm all beings. To date there are six declarations that have documented key moments in the Zapatista struggle for autonomy. The first one in January 1994 declared war on the Mexican government, and the Sixth Declaration in June 2005 announced a complete disengagement with the Mexican State and a focus on building autonomous governance structures. The Sixth, as this last declaration is commonly known, narrates how their autonomous forms of government came into being as a way to govern from below. It is also a call for 'an international' below, for all who have been marginalized by globalization from above, to commit to the struggles in their own geography, in their own way.

Nuclear Colonialism

I first learned about the properties of uranium (U) while researching the use of depleted uranium weapons by the United States during the 1990–91 US-lead war on Iraq, known as the Gulf War. Uranium, a radioactive heavy metal element used as fuel for nuclear power plants and nuclear weapons, is mined for extraction from ore deposits. Uranium is radioactive, which means it is an unstable element that emits energy through a process of decay. This decay or energy loss takes place through the emission of radiation in the form of alpha, beta and gamma rays. As uranium decays, it transforms into other different unstable elements through a decay chain until it turns into the stable element of lead (Pb). Each decay process has its own half-life, the time it takes for radioactivity to decrease to half of its original value. The radioactive and chemical properties vary with each element in the decay chain. The half-life of uranium is 4.5 billion years. It is part of the earth's crust, and in fact the date of the Earth is in part determined by the uranium decay timeline.

Uranium ore deposits in the earth's crust contain different isotopes, variations of a given element that contain the same number of protons but a different number of neutrons. The most common uranium isotopes are U-238 and U-235. The uranium enrichment process increases the concentration of U-235, the isotope used for nuclear fission needed in nuclear fuel and weapons. The waste byproduct of this enrichment process is depleted uranium, and while it contains less of a U-235 concentration, depleted uranium is nonetheless radioactive.[5] Because of its extremely high density, depleted uranium is used on projectile weapons in order to pierce tanks and also for the same reason used to armour-plate tanks for reinforcement. Depleted uranium does not activate nuclear fission but is used at the penetrating tip of a projectile to pierce through armours and burn upon impact, releasing radioactive and chemically toxic dust. The early medical research on the health effects of exposure to depleted uranium during the Gulf War drew from studies of those who were exposed to radiation from uranium mining because this was a problem that had been studied for a longer period and could be helpful given that symptoms from radiation exposure do not show up for many years (Institute of Medicine 2008). This is when I started to learn about uranium mining, the concept of nuclear colonialism and the peculiar role of New Mexico.

New Mexico is home to Los Alamos National Laboratory, where the United States government secretly operated the Manhattan Project for the development of the nuclear bomb during the Second World War. Uranium mining began, also secretly, in 1941 under the Army Corps of Engineers' Manhattan Project for the creation of the nuclear bomb. The US government first tested an atomic bomb in the Jornada del Muerto (Journey of Death) Valley in New Mexico (code named Trinity) and in the Bikini Atoll coral reef in the Marshall Islands before exploding them above the more densely populated cities of Hiroshima and Nagasaki in August of 1945, killing over 200,000 people at the time and more later due to radiation exposure. The following year, the newly formed US Atomic Energy Commission took over the task of overseeing the build-up of atomic science and technology and started a uranium ore buying programme in 1948 (Rautman 1977). The state of New Mexico was an intensive site of uranium mining and milling from the 1940s until the 1980s, when most uranium mining halted. Nuclear deterrence was the logic of the Cold War, and with the escalation of the nuclear arsenal, the market for uranium mining was robust. The uranium-rich Grants Mining District in northwestern New Mexico, stretching between the cities of Albuquerque and Grants, crosses Diné and other tribal lands. There are currently upwards of 500 abandoned open pit uranium mines in Diné territory alone, and some estimates go as high as 2000. The abandoned Jackpile-Paguate Uranium mine located in Laguna Pueblo was operative from 1952 through 1982, during which about 400 million tons of rocks were moved within the mine area and 25 million tons of uranium ore were transported out. Radon gas from the mine dust contaminated the area, and people who worked in the uranium mines continue to suffer from different kinds of cancer. There are many more examples.

In addition to struggles to contain radioactive contamination from abandoned mines, activists are engaged on multiple fronts to prevent new uranium mining as well as proposed radioactive waste storage projects. I first learned that New Mexico is connected to almost every single aspect of the nuclear fuel chain from one of

Leona Morgan's presentations. Morgan is co-founder of the Nuclear Issues Study Group and a Diné community organizer and activist, who offers public talks and webinars on the history and politics of the nuclear fuel chain and advocates for communities affected by nuclear colonialism.[6] She speaks about New Mexico as the birthplace of nuclear colonialism and describes the uneven distribution of risk and exposure affecting Native communities and the logic of New Mexico as a national sacrifice zone. She visually marks the multiple locations of radioactive projects on the New Mexico state map. The Trinity site of the atom bomb explosion is probably the most well known. The Grant Mineral Belt is riddled with abandoned open-pit uranium mines that continue to pollute the air, soil and waterways. The Waste Isolation Pilot Plant near Carlsbad is an underground facility for the deep geological disposal of the entire nation's spent nuclear fuel from nuclear power plants and high-level radioactive waste, run by the US Department of Energy. Holtec International, a private company based in New Jersey, has applied for another site as a Consolidated Interim Storage (CIS) facility for commercial high-level radioactive waste in southeast New Mexico. There is even depleted uranium weapons testing at White Sands Missile Range, Los Alamos National Laboratories, Sandia National Laboratories in Albuquerque, and facilities with New Mexico Tech in Socorro.

While there is comparatively little uranium mining at the moment, companies continue prospecting new and old sites,[7] and the industry has been lobbying the United States government to support domestic uranium production in order to not rely on foreign uranium for reasons of national security. The argument has been updated to emphasize the further insecurity of the supply chain since the Covid-19 pandemic (National Mining Association 2020). The nuclear industry proposes the new mining technology of In Situ Leach (ISL) mining as a safe alternative to the previously practised open-pit mining. The Black Hills Clean Water Alliance (South Dakota) reports that ISL has to be done in water-bearing aquifers, which is the source of drinking water.

> In situ leach mining (ISL) is a method of uranium mining where hundreds of wells are drilled in a 'grid pattern' over an ore body that is located in a groundwater aquifer. Water mixed with sodium bicarbonate concentrate is pumped down into the aquifer, across the uranium ore bed, and then up and out other wells. The mining solution is injected into the aquifer *under pressure* in order to leach the uranium out of the ground. The leach solution strips the uranium out of the aquifer. Along with radioactive uranium, *arsenic, selenium, radium and lead* are also extracted. (BHCWA 2017)

Although Diné/Navajo Nation banned uranium mining and milling in 2005 and the transport of radioactive materials in 2012, the governing of these activities remains contested.

For instance, the indigenous-led group HAUL NO! has been organizing direct actions and resistance against Energy Fuel Inc's plans to transport uranium from Pinyon Plain Mine (previously known as Canyon Uranium Mine) in the state of Arizona to White Mesa Uranium Mill close to the Ute tribal community in the state of Utah.[8] The proposed route, which runs in part through Diné territory, stretches across 483 km of Native lands and populated cities, but tribal governments do not have jurisdiction over state and federally controlled highways. Tribal lands and private allotments intermingle with state and federal lands. Currently, a highly contested uranium mining project at Crownpoint, New Mexico sits a few kilometres from the site of the 1979 radiation spill and threatens to contaminate the same region.

The breach at the United Nuclear Corporation uranium mill in Church Rock that Cody described in her opening remarks released more radiation than the well-known Three Mile Island nuclear power plant accident, which had taken place in Pennsylvania only four months before. The Church Rock spill remains largely unknown despite its greater radioactive magnitude.[9] The lack of national response to the radioactive spill appears as a way to downplay the actual harm – *yes, there was a spill, but there are no real threats to public health.* Such disavowal of life can be traced to the beginnings of colonial experience in the Americas. A way to understand the national insignificance of the largest national radiation accident is to situate the Church Rock uranium mill spill within the colonial legacy of declaring a territory *terra nullius*, empty land. More precisely, it meant land belonging to no one. The Latin term was first used for land that did not belong to the Roman Empire and later in the Middle Ages for land that did not belong to a Christian ruler. It was a term used in international law to legalize the conquest of land and its inhabitants. *Terra nullius* is the older more general term used by European colonial powers since the Treaty of Tordesillas. In the United States, this concept and practice has been articulated through the Doctrine of Discovery and the idea of Manifest Destiny. Empty land but because it was not empty an argument had to be made that the land is empty. Erasure and also disavowal: I know there are humans here, but nonetheless they are not human.

Dislocated Events

Nuclear colonialism is connected to other practices of domination described as extractivism, resource colonialism, environmental racism, racial capitalism, militarism, nationalism, imperialism, patriarchy. There are other helpful terms. The Zapatistas offer the image of a Capitalist Hydra to describe the current world-system, a many-headed monster capable of regenerating its multiple heads, agile, cunning and able to shapeshift. They nod to the debates about the arrangement of the heads – is the analysis of the State more important or patriarchy, for example? But they move on from such rankings to ask how we might go about defeating the Hydra. They propose a seminario/semillero, a seminar for exchanging and generating a seedbed of ideas, because we need everyone's critical thinking, each from their own different geographies, to help face the many shapes, forms and heads of the changing Hydra. They describe this seedbed seminar as a 'dislocated event' that will take place across different calendars and geographies.

> That's why we say that it's a 'dislocated' event, that it doesn't happen in just one place but in many places, all over the place. And we say that it is worldwide because there is critical thinking in all of the worlds that there are out there, that everywhere people are inquiring about what is going on, why it's happening, what to do, how to do it, and all of these questions that are thought out through theory. (Sixth Commission 2016: 20)

The International Uranium Film Festival itself felt like such a dislocated event, where people from many different places shared their experiences and knowledges from below. Twenty-two films screened at Window Rock and ranged in their geographic focus from Fukushima to Cree and Diné territory, Ute tribe in Utah, the Marshall Islands, Greenland, India and Australia – this is not an exhaustive list. For three days, we were immersed in films and discussions of systematic state, military and corporate strategies of violence, abandonment and secrecy, which continue to undermine basic life-chances – access to uncontaminated water, air, soil and food ways. One particular post-screening discussion turned into information sharing about known remedies and practices for responding to radiation exposure, such as ingesting bentonite clay as a detoxing agent, and how apple pectin was used after Chernobyl to help eliminate radioactive particles from intestinal tracts. During private conversations, there were also stories of disagreements around organizing, Native activists in conflict with capitalist and/or corrupt tribal governments, misunderstandings over funding, time and labour, unkept promises, genuine mistakes, and cultural inaccuracies in a film.

The film festival at Window Rock was one in a series that stretched across several towns and cities. Screenings were held in Window Rock, Flagstaff, Albuquerque, Grants, Santa Fe and Tucson. In this sequence, the screenings did not follow a straight line; instead, they went back and forth, retracing paths across a territory marked by conquest, forced displacements, capitalist state building, mapping, treaty agreements, prospecting, intensive mining and also peach orchards, pastures, rebellions and resistances, mutual-aid, and sacred mountains (Brugge, Benally and Yazzie-Lewis 2006; Dunbar-Ortiz 2007; Powell 2018; Voyles 2015). These are interconnected histories and not discrete items in a list that might imply comparable events or practices (or even units of grammar). *There are also different kinds of differences.* For example, sixteenth-century Spanish conquest was different from the Anglo-American nation state building. Yet both shared a similar colonial framework that in the aftermath of a series of papal bulls justified the violent and bloody seizure of land from non-Christian inhabitants, whose humanity would be a matter of recorded dispute (Wynter 2003). Specifically, the land tenure system under Spanish colonial rule in New Mexico included some forms of communal ownership of land and cooperative management of intricate water canals (*acequias*) to irrigate crops such as corn, beans and squash. With the United States government incorporating New Mexico into its southern borders after the Mexican-American War, there was a different practice of expropriation of communal lands, and a shift from trade-based mercantile capitalism to the commodification and privatization of the land. By the mid-twentieth century, New Mexico's economy was integrated within the national capitalist economy through the nuclear industry and uranium mining (Dunbar-Ortiz 2007: 123).

Patrick Wolfe's (2006) insight that settler colonial 'invasion is a structure not an event' suggests that colonial legacies are not bound by the past but continue to be rearticulated through existing systems and practices. The logic of the 'elimination of the native', to continue with Wolfe's language, *attempts* to eliminate Native lives through war or assimilation by erasing ways of living, relations to land, kinship and governance. Critical instances include the Native American land reservation system created in 1851 by the United States government to confine Native American populations to distinct areas and make more land available for settlers. The General Allotment Act of 1887, also known as the Dawes act, imposed the breaking up of communal tribal lands into individual plots designed to dismantle communal forms of living, belonging and governance and to assimilate Native populations into a capitalist economy. Refusal to submit meant not belonging to an official tribal roll created through this very process and being denied US citizenship. The imposition of a nuclear family structure was tied to individual plots of land as private property owned by men as the legitimate 'head' of the family. With the ownership of land made available exclusively for men, matrilineal belonging was displaced. Instead, a racialized concept of tribal belonging (TallBear 2013) came into being, and Native populations were dispossessed of millions of acres of tribal land after the allotments. Decades later, with the Indian Reorganization Act of 1934, the US federal government recognized the newly imposed 'tribal council' governments on reservations as the only legitimate body with which it would negotiate such things as land transfers, water rights and mineral leases. More to the point, such tribal governments were put into place precisely to facilitate private, settler state and corporate access to Native lands (Brugge, Benally and Yazzie-Lewis 2006).

Spanish and US colonialism differed then in their habits of conquest. With the emergence of the US capitalist state, a more intensified practice of private accumulation of land by dispossession of Native communal life-land relations was practised illegally and legally, through laws created for that very purpose. *A different kind of difference* from the difference between Spanish and US land relations marks Native/Indigenous life-land relations. Human Rights Lawyer June Lorenzo writes about how for her Laguna Pueblo peoples and for other Indigenous peoples, the land is experienced as a relative, a living being, and that land stewardship and not ownership orients this relation (2017). Similarly, the words of Diné activist, musician and filmmaker Klee Benally's grandmother Roberta Blackgoat:

> The Coal they strip mine is the earth's liver. The earth's internal organs are dug up. Mother Earth must sit down. The uranium they dug up for energy was her lungs. Her heart and her organs are dug up because of greed. It is smog on the horizons. Her breath, her warmth, is polluted now and she is angry when Navajos talk of their sickness. The coal dust in winter blows in to blanket the land like a god down the canyons. It is very painful to the lungs when you catch a cold. The symptoms go away slowly when dry coal dust blows in from strip mining. The people say the uranium can dry up your heart. No compassion is left for the motherland. We've become her enemy. (Benally 2020)

In the absence of a shared understanding of the Earth as a living being by the United States, Native activists strategically use the

US legal system to protect the Earth. The Hopi, Zuni, Laguna and Acoma, for example, have successfully fought to protect the sacred site of Mt. Taylor (northwest New Mexico) from uranium and other forms of mining through the legal system under the New Mexico Cultural Properties Act (Lorenzo 2017). This sacred relation to the land and its elemental beings was palpable across many of the films screened and conversations at the festival in Window Rock.

The very first International Uranium Film Festival took place in 2011 in Rio de Janeiro by co-founders and co-directors Marcia Gomes de Oliveira and Norbert Suchanek as an annual event with additional travelling screenings around the world, including Germany, Japan, Jordan, Palestine, Canada and others. The International Uranium Film Festival has an oddly interesting story of origin.[10] In 2006 Suchanek attended the Indigenous World Uranium Summit in Window Rock, carrying with him a video message from Chief Verá Mirim of the Sapukai village of the Guarani Mbyá indigenous community. As it happens, the Brazilian nation state bureaucracy and its relation to indigenous communities was such that Chief Verá Mirim was not able to secure the necessary paperwork for travel to the United States in time to attend the Summit. Suchanek, being a German journalist living in Brazil, was able to attend. He scoffs at the absurdity of a Bavarian *representing* the indigenous community of the Sapukai village at the Indigenous World Uranium Summit. He sighs and we all chuckle. Pause. Perhaps to mitigate the absurdity, Gomes de Oliveira and Suchanek tell me of their deeper history with the tribe and how they had been married by another chief in the village. It's an intimate and beautiful story of seemingly unusual connections, pacts and promises. They then continue with the account of the foiled travel plans and how the day before Suchanek's flight they filmed an interview with Chief Verá Mirim as a way to deliver a direct message from him to the Indigenous World Uranium Summit:

> My village here is a big community. I have many grandchildren. Many families. This is why I am very worried about the Nuclear Power Plants. Because my village is close to them. Very close! This is why I am concerned. There is big danger, big risk. Perhaps there is a leak. It happens an accident with the Nuclear Power Plants. My village will be affected. Clearly my village will be affected! Very certainly! Because it is very close to the power plants.[11]

Chief Verá Mirim's palpable distress for the future of his people and for his grandchildren anticipates the generational inheritances of radiation from a possible nuclear accident. Suchanek explains to me that Chief Verá Mirim is particularly worried about a possible nuclear accident because there is no existing procedure for evacuating his community situated in the mountains of the Atlantic rainforest. 'The only access to his village is a difficult dirt track that turns into a mud gully in the rainy season and is almost impossible to cross, even with a four-wheeler.' He further emphasizes the proximity of the two nuclear power plants Angra 1 and Angra 2, with a third Angra 3 under construction at the same site, Almirante Álvaro Alberto Nuclear Power Center (CNAAA).

Suchanek successfully carried Chief Verá Mirim's message not by standing in for him but rather by facilitating his participation. The idea for the International Uranium Film Festival emerged then, at the 2006 Indigenous World Uranium Summit, where two older documentaries about environmental racism had made an impact – one about the long-term health and environmental consequences of uranium mining on the Diné in Monument Valley (*The Return of Navajo Boy*, 2001), the other about tribal Adivasi communities in Jadugoda, India, where rice fields are contaminated by the dumping of uranium waste byproducts from a nearby uranium mine (*Buddha Weeps in Jadugoda*, 1999). Struck by the fact that he had not encountered these important films that had been made years prior, Suchanek was encouraged to create an international film festival dedicated to bringing more awareness to the subject matter of uranium.

Gomes de Oliveira was hesitant at first. She recalls the moment she changed her mind about four years later, in 2010, when a newspaper headline in Brazil announced the country's stated goal to be the number one exporter of uranium in the world. 'But most people didn't even know what is *urânio*! Maybe they think it's a star! People don't know the dangers or even that there is a relation between uranium and nuclear.' She throws her hands up exasperated, which makes a strong impression given her petite frame and soft eyes. In Portuguese, *urânio* is uranium and Urano is Uranus, the seventh planetary body from the Sun after which Martin Heinrich Klaproth named the element in 1789. She spoke passionately about how important it is for people to know about uranium and the dangers in unearthing it. The role of educating as many people as possible about nuclear issues is, not surprisingly, a widely shared sentiment amongst anti-nuclear activists given the level of secrecy around all things nuclear. But Gomes de Oliveira does not consider herself or the festival itself 'anti-nuclear', and wants to build relationships internationally, share knowledge and exchange stories with as many people as possible. Suchanek and Gomes de Oliveira had noticed that those who attend the anti-nuclear events are usually always the same people. That there is no engagement with a wider public. Their aim was to break through 'this invisible wall' with the film festival.

The 2018 Window Rock event was then a poignant moment of return to the birthplace of the festival twelve years prior. It was my first time attending the festival after following their online archive of international screenings for some time. All the local activists I had met during previous research trips to New Mexico were present. There was a general sense of affinity and friendliness in the audience, not entirely without tensions, which I would learn about later. It was not until Cody's reference to the Zapatistas on the morning of the second day that I felt a sense of ease wash over me. 'Everything for everyone and nothing for ourselves.' I was affectively transported to the autonomous Zapatista spaces I had been participating in during my annual trips to Chiapas since 2012. My experience of these autonomous spaces feel quite different from those of the friendliest activist and academic spaces. The difference is clear, deeply palpable and difficult to describe; a sense of reciprocity and mutuality overflows neoliberal individual subjectivities that cultivate competitive self-interest despite our best intentions otherwise. Could this be a 'dislocated event' for genuine sharing of experience, knowledge and critical thinking from different parts of the world, to respond to Sabu Kohso's insightful observation that 'the strongest weapon of the nuclear regime is this incapacitation of our thinking and acting' (2020: 9)? Was I romanticizing the film festival? Perhaps. The Zapatistas? I hope not.

Conclusion

In order to situate uranium mining in New Mexico within the global nuclear fuel chain and nuclear colonialism in the space of this chapter, I have risked offering only a brief sketch of the many and calamitous harms of radiation exposure and chemical toxicity many peoples have experienced because of uranium mining in New Mexico (Brugge, Benally and Yazzie-Lewis 2006; Voyles 2015; Pasternak 2010; Powell 2018).

I continue to learn from fierce local activists, whose lifework has involved community organizing, strategic legal battles with the state, epidemiological studies, direct action and mutual aid.[12] I chose to give space to the possibilities of connecting different struggles and resistances to nuclear colonialism from across different geographies by focusing on the example of the International Uranium Film

Festival as a site of exchange from below because we need to learn from each other.

The Danish artist and filmmaker Lise Autogena was in attendance to screen and discuss her film about uranium prospecting in the region of Kvanefjeld in southern Greenland, home to the mostly indigenous community of Narsaq, and one of the world's largest deposits of uranium. A former official colony of Denmark, Greenland is currently an autonomous constituent country of the Kingdom of Denmark and still struggling for independence. The film *Kuannersuit; Kvanefjeld* (2017) is entirely comprised of conversations with a range of community members, from politicians to farmers and teachers, without a narrative voiceover to guide a particular perspective. It reveals the contradictions facing a deeply divided community eager to gain financial independence from Denmark and disengage from its colonial past. The melting ice sheet in the region is opening up opportunities for mining, which for a population desperate for independence and in need of substantial capital represents a means of gaining autonomy and financial independence. What could be one of the world's largest uranium mines would in effect destroy their existing ways of living from the land and the sea. During the post-screening discussion, some suggested that films screened at Window Rock be shared with the Narsaq community to show the inevitability of radiation contamination and the impossibility of returning a local ecology to its previous state so they can know in advance the full consequences of their decision. It was clear from the film that the residents were not fully informed about the consequences of living 6 km from the planned mine. I do not want to downplay the affective dimension of the shared experiences palpable during our screenings and discussions, the realization that one's plight is not exceptional, that communities that otherwise might feel isolated and abandoned are in fact not alone and there are others who are fighting similar battles, that they are related to each other and that they can be in relations of mutual aid.

This type of pattern recognition also helps us see the structural vulnerability (Quesada, Hart and Bourgois 2011) or existing structures of dispossession that makes nuclear colonialism possible. According to a recent report, Covid-19 patients 'in areas impacted by high levels of air pollution before the pandemic are more likely to die from the virus than patients in other parts of the US' (Benally 2020). The lack of access to uncontaminated water for the Diné is another long-term problem that has been further intensified by the pandemic. The unambiguous pattern actively suppressed by power from above is that there has never been a successful 'clean up' of radioactive contamination from conventional open pit-mining, and despite the assurances that the technically more advanced In Situ mining is safer, there is strong evidence to the contrary. 'To date, no remediation of an ISR operation in the United States has successfully returned the aquifer to baseline conditions' (Hall and Otton 2009). Even where such scientific studies exist, they might not readily be available to different communities in the process of being courted. Another insight shared was that the most effective way to prevent future uranium mining is to interfere at the stage when companies are trying to secure permits for prospecting. This strategy might not work in every context, given, for example, that in India there is no distinction between mining companies and the government, which might feign oversight and a system of accountability. This point about the particular difficulty of addressing harm from uranium mining in India because the Uranium Corporation of India operates under the government was made by activist filmmaker Shriprakash.

As mentioned earlier, Shriprakash's documentary (*Footprints*, 2018) about the plight of the Adivasi indigenous communities, whose food and waterways have been contaminated by the dumping of radioactive waste from uranium mining, was screened at the 2006 Indigenous World Uranium Summit in Window Rock and was an inspiration for creating the film festival. After attending the Summit and learning about the local struggles there, Shriprakash was in turn moved to make another documentary connecting the struggles of the Acoma, Laguna and Diné peoples to the Adivasi experience in India. There are other long-term connections that have emerged from these spaces of encounter.

We might say that the shared experiences of different peoples and geographies due to nuclear colonialism is only one way to frame how struggles are interconnected under the dominant capitalist worldview today. There are other worldviews that hold all life on Earth as always already interconnected. My grandmother's view of the world through the poetry of Sa'di and Cody's grandmother's life practice of K'é share a sensibility of the interconnected web of life, even if they are from vastly different worldviews.

Throughout this chapter, I have attempted to highlight the knowledges and connections from below even if at times the range and juxtaposition of references seemed awkward. I want to end by also recognizing the important role that scientists play in mitigating the harm of the global nuclear fuel chain. In *Journey to the Safest Place on Earth* (2013), we learn how nuclear waste storage vessels degrade.

The documentary follows nuclear physicist Charles McCombie in his search for a permanent underground location to safely store over (at the time) 350,000 of tons of existing spent nuclear fuel from power plants in temporary storages that are breaking down around the world. The technology needed to safely contain the buried radioactive spent fuel, which will remain toxic for at least a 100,000 years (or more), does not exist. This stark picture of reality should provide a strong case for ending all uranium mining as it is the source point for the entire nuclear fuel chain. It also makes the case for the expertise of nuclear physicists and scientific technology to mitigate existing threats.

Some scientific ways of thinking and acting in the world strive for human superiority over nature as an object to be studied and controlled and its perceived threats overcome. At least this is the image of the world that has dominated the capitalist world-system. But this is not the only vision of the world. There are other scientists for whom the Earth as a living organism offers a more explicit framework for understanding and participating with our shared world as part of the interconnectedness of all life (Barad 2007; Cajete 2000; Kimmerer 2015; Lovelock and Margulis 1974). In fact the sub-atomic world might offer us a theory of a future that can overcome our seemingly intractable trajectory, and physicist Karen Barad's words seem like a good place to end for now:

> Future moments don't follow present ones like beads on a string. Effect does not follow cause hand over fist, transferring the momentum of our actions from one individual to the next like the balls on a billiards table. Causality is an entangled affair: it is a matter of cutting things together and apart (within and as part of phenomena). It is not about momentum transfer among individual events or beings. The future is not the end point of a set of branching chain reactions; it is a cascade experiment. (Barad 2007: 394)

Targol Mesbah is Associate Professor in Anthropology and Social Change at the California Institute of Integral Studies in San Francisco. She researches the logistics and afterlives of wars and experiments with how the Zapatista concept of history as a network of 'calendars and geographies from below' can guide research whose temporality and ethics are designed for life-sustaining practices.

Notes

1. Navajo is the English term for Diné. The Navajo Nation is federally recognized and the largest Native American tribal land reservation in the United States. Within the legal framework of the federal government, Native tribes have limited sovereignty and self-determination, and Native reservations are held in trust by the Department of Interior and ultimately owned by the US federal government. On the politics, contradictions and different notions of sovereignty, see Kauanui (2018: 25–28).
2. The video of Cody's opening remarks is available on the Facebook page for the International Uranium Film Festival. https://www.facebook.com/uraniumfilmfest/videos/2218476998390955. For more on the concept of K'é, see K'é Infoshop (n.d.) and Denetdale (2020).
3. The expression 'nuclear colonialism' was used by Valerie L. Kuletz (1998) in her account of the nuclear contamination in the American West and echoes the earlier use of the formulation of 'radioactive colonialism' by Ward Churchill and Winona LaDuke (1986).
4. The Navajo Birth Cohort Study, a collaborative research project and the first epidemiological study of the relationship between uranium exposures and reproductive health, tracks birth defects and developmental delays in children of families exposed to the mutagenic properties of uranium in the Diné. See https://hsc.unm.edu/pharmacy/research/areas/nbcs-echo.html.
5. Gabrile Hecht (2012) has written extensively on the global politics of how and when things are marked as being nuclear.
6. The Radiation Monitoring Project, which Morgan is involved with, provides communities with access to technology and training in order to monitor radiation levels in their area. See http://www.dinenonukes.org/radiation-monitoring-project/.
7. See the World Nuclear Association website for updates and tracking of uranium mining activities. https://world-nuclear.org/information-library/country-profiles/countries-t-z/us-uranium-mining.aspx.
8. See the website of HAUL NO! at https://haulno.com/.
9. The Three Mile Island nuclear power plant in Pennsylvania released 13 curies of nuclear fission byproducts, whereas the Church Rock Mill spill released 46 curies of transuranic isotopes and heavy metals (Brugge, DeLemos and Bui 2007).
10. Personal communication.
11. Suchanek shared the transcript of the Chief's video through email correspondence.
12. In the intervening months during this book's editorial process, Klee Benally tragically passed away shortly after the release of his book *No Spiritual Surrender: Indigenous Anarchy in Defense of the Sacred* (2023). He shares his journey and analysis as a Diné in struggle against settler colonial violences of the sacred while his uncompromising prose unravels for us any liberal cooptation of the sacred. Projects he was deeply involved with include Indigenous Action at https://www.indigenousaction.org/ and Haul No! https://haulno.com/.

References

Barad, Karen. 2007. *Meeting the Universe Halfway: Quantum Physics and the Entanglement of Matter and Meaning*. Durham, NC: Duke University Press.

Benally, Klee. 2020. 'Covid-19, Resource Colonialism & Indigenous Resistance'. Indigenous Action. Retrieved 2 November 2023 from https://www.indigenousaction.org/covid-19-resource-colonialism-indigenous-resistance/.

______. 2023. *No Spiritual Surrender: Indigenous Anarchy in Defense of the Sacred*. Olympia, WA: Detritus Books.

BHCWA. 2017. *What is 'In Situ' Mining?* Black Hills Clean Water Alliance. Retrieved 2 November 2023 from https://bhcleanwateralliance.org/what-is-in-situ-mining/.

Brugge, Doug, and Rob Goble. 2002. 'The History of Uranium Mining and the Navajo People', *American Journal of Public Health* 92(9): 1410–19.

Brugge, Doug, Timothy Benally and Esther Yazzie-Lewis. 2006. *The Navajo People and Uranium Mining*. Albuquerque: University of New Mexico Press.

Brugge, Doug, Jamie L. deLemos and Cat Bui. 2007. 'The Sequoyah Corporation Fuels Release and the Church Rock Spill: Unpublicized Nuclear Releases in American Indian Communities', *American Journal of Public Health* 97(9): 1595–600. https://doi.org/10.2105/AJPH.2006.103044.

Buddha Weeps in Jadugoda. 1999. Directed by Shriprakash [Video]. Jadugoda: Kritika & Birsa.

Cajete, Gregory. 2000. *Native Science: Natural Laws of Interdependence*. Santa Fe: Clear Light Publishers.

Churchill, Ward, and Winona LaDuke. 1986. 'Native America: The Political Economy of Radioactive Colonialism', *Insurgent Sociologist* 13(3): 51–78.

Comité Clandestino Revolucionario Indígena-Comandancia General del Ejército Zapatista de Liberación Nacional. 1994a. *Segunda Declaración de la Selva Lacandona*. Retrieved from https://enlacezapatista.ezln.org.mx/1994/06/10/segunda-declaracion-de-la-selva-lacandona/.

______. 1994b. *Sobre la consulta: para todos todo, nada para nosotros*. Retrieved from https://enlacezapatista.ezln.org.mx/1994/06/10/sobre-la-consulta-para-todos-todo-nada-para-nosotros/.

Cusicanqui, Silvia Rivera. 2020. *Ch'ixinakax utxiwa: On Practices and Discourses of Decolonization*. Cambridge: Polity Press.

Denetdale, Jennifer Nez. 2007. *Reclaiming Dine History: The Legacies of Navajo Chief Manuelito and Juanita*. Tucson, AZ: University of Arizona Press.

______. 2020. 'Refusing the Gift of Democracy and Embracing Diné Concepts of Kinship: The Navajo Nation, Citizenship, and Practices of Gender', *Theory & Event* 23(4): 1053–65.

Dunbar-Ortiz, Roxanne. 2007. *Roots of Resistance: A History of Land Tenure in New Mexico*. Norman, OK: University of Oklahoma Press.

Footprints. 2018. Directed by Shriprakash [Video]. Jadugoda.

Hall, S., and J.K. Otton. 2009. 'In-situ Recovery Uranium Mining in the United States: Overview of Production and Remediation Issues'. *International Symposium on Uranium Raw Material for the Nuclear Fuel Cycle: Exploration, Mining, Production, Supply and Demand, Economics and Environmental Issues*. Retrieved from https://inis.iaea.org/search/search.aspx?orig_q=RN:41003255.

Hecht, Gabrielle. 2012. *Being Nuclear: Africans and the Global Uranium Trade*. Cambridge, MA: MIT Press.

Hinchman, Steve. 1987. 'Rebottling the Nuclear Genie', *Hight Country News*, 19 January, 19: 1, Front page.

Hindin, Rita, Doug Brugge and Bindu Panikkar. 2005. 'Teratogenicity of Depleted Uranium Aerosols: A Review from an Epidemiological Perspective', *Environ Health* 4(17). DOI:10.1186/1476-069X-4-17.

Institute of Medicine (US) Committee on Gulf War and Health. 2008. 'Updated Literature Review of Depleted Uranium', in *Epidemiologic Studies of Veterans Exposed to Depleted Uranium: Feasibility and Design Issues*. Washington D.C.: National Academics Press.

Journey to the Safest Place on Earth. 2013. Directed by Edgar Hagen [Video]. Switzerland.

Kauanui, Kēhaulani J. 2018. *Paradoxes of Hawaiian Sovereignty*. Durham, NC: Duke University Press.

———. 2021. 'The Politics of Indigeneity, Anarchist Praxis, and Decolonization', *Anarchist Developments in Cultural Studies* 2021(1): 9–42.

K'é Infoshop. n.d. *Settler Sexuality: Resistance to State-Sanctioned Violence, Reclamation of Anti-Colonial Knowledge and Liberation for All*. An Indigenous Feminist Zine. http://keinfoshop.org/zines/settler-sexuality.htm.

Kimmerer, Robin Wall. 2015. *Braiding Sweetgrass: Indigenous Wisdom, Scientific Knowledge, and the Teaching of Plants*. Minneapolis, MN: Milkweed Editions.

Kohso, Sabu. 2020. *Radiation and Revolution*. Durham, NC: Duke University Press.

Kuannersuit; Kvanefjeld. 2017. Directed by Lise Autogena and Joshua Portway [Video]. UK.

Kuletz, Valerie L. 1998. *The Tainted Desert: Environmental Ruin in the American West*. New York: Routledge.

Lorenzo, June L. 2017. 'Spatial Justice and Indigenous Peoples' Protection of Sacred Places: Adding Indigenous Dimensions to the Conversation' ['Justice spatiale et protection des lieux sacrés par les peuples autochtones : intégrer des dimensions autochtones à la discussion', traduction : Carole Cancel]. *justice spatiale | spatial justice* 11. http://www.jssj.org.

Lovelock, James E., and Lynn Margulis. 1974. 'Atmospheric Homeostasis By and For the Biosphere: The Gaia Hypothesis', *Tellus* 26(1–2): 2–10.

Margulis, Lynn. 2008. *Symbiotic Planet: A New Look at Evolution*. New York: Basic Books.

Masco, Joseph. 2006. *The Nuclear Borderlands: The Manhattan Project in Post-Cold War New Mexico*. New Jersey: Princeton University Press.

Morizumi, Takashi. 2002. *A Different Nuclear War: Children of the Gulf War*. Hiroshima: Association for Banning Depleted Uranium Weapons.

National Mining Association. 2020. 'Establishment of National Uranium Reserve Key to Support U.S. National Security, Secure Our Supply Chains'. Press Release. Retrieved from https://nma.org/2020/04/23/establishment-of-national-uranium-reserve-key-to-support-u-s-national-security-secure-our-supply-chains/.

Nixon, Rob. 2011. *Slow Violence and the Environmentalism of the Poor*. Cambridge, MA: Harvard University Press.

Pasternak, Judy. 2010. *Yellow Dirt: An American Story of a Poisoned Land and a People Betrayed*. New York: Free Press.

Physicians Task Force on the Health Risks of Nuclear Weapons Production. 1992. *Dead Reckoning: A Critical Review of the Department of Energy's Epidemiologic Research*. Washington D.C.: Physicians for Social Responsibility.

Powell, Dana E. 2018. *Landscapes of Power: Politics of Energy in the Navajo Nation*. Durham, NC: Duke University Press.

Quesada, James, Laurie Kain Hart and Philippe Bourgois. 2011. 'Structural Vulnerability and Health: Latino Migrant Laborers in the United States', *Medical Anthropology* 30(4): 339–62. DOI: 10.1080/01459740.2011.576725.

Rautman, Christopher. 1977. *The Uranium Industry in New Mexico*. Open File Report 74. Socorro: New Mexico Bureau of Mines and Mineral Resources.

Scarry, Elaine. 2014. *Thermonuclear Monarchy: Choosing Between Democracy and Doomsday*. New York: W.W. Norton & Company.

Shuey, Chris. 1981. 'Calamity at Church Rock', *Scottsdale Daily Progress*, 14 and 21 February and 21: Saturday Magazine: 3–6.

Sittnick, Philip. 1998. *Uranium Mining and its Impact on Laguna Pueblo: A Study Guide for an Interdisciplinary Unit*. For Laguna Middle School.

Sixth Commission of the Zapatista Army of National Liberation. 2016. *Critical Thought in the Face of the Capitalist Hydra*. Durham, NC: Paperboat Press.

TallBear, Kim. 2013. *Native American DNA: Tribal Belonging and the False Promise of Genetic Science*. Minneapolis, MN: University of Minnesota Press.

The Return of Navajo Boy. 2001. Directed by Jeff Spitz. Berkeley, CA: Berkeley Media.

Tsosie, Rebecca. 2015. 'Indigenous Peoples and the Ethics of Remediation: Redressing the Legacy of Radioactive Contamination for Native Peoples and Native Lands', 13 Santa Clara J. Int'l L. 203.
Voyles, Traci Brynne. 2015. *Wastelanding: Legacies of Uranium Mining in Navajo Country*. Minneapolis, MN: University of Minnesota Press.
Wolfe, Patrick. 2006. 'Settler Colonialism and the Elimination of the Native', *Journal of Genocide Research* 8(4): 387–409.
Wynter, Sylvia. 2003. 'Unsettling the Coloniality of Being/Power/Truth/Freedom: Towards the Human, After Man, Its Overrepresentation – An Argument', *CR: The New Centennial Review* 3(3): 257–337.
Yellow Fever: Uncovering the Navajo Uranium Legacy. 2013. Directed by Sophie Rousmaniere [Video]. Austin: Issue TV.

— Afterword —

What's to Come

Mining in a Fevered World

David A. Kideckel

The world is burning up, and global mining has long been the arsonists' willing accomplice – often even the firestarter. Pusceddu and Zerilli's introduction to the global life of mines sharply portrays how mining in the global context and over the centuries has produced the forces behind its incendiary qualities. As they relate, the rapacious hunt for mineral wealth and the astonishing inequalities and spreading environmental and cultural transformation this enables position mining as a key force in the cataclysmic shaping of the globe and its peoples, especially since the advent of global capitalism.

The chapters in this volume fit squarely into this discussion. Below I discuss both their political-economic bent and also what they imply for future directions in the study of mining and post-mining. Despite their geographic and thematic disparity, the chapters have a singular focus. They trace how global mining enables that same inequality in land, challenged identity and threatened health and life that are part of extraction's arsonist bent. At the same time, the chapters, dire as they often are, offer hope, as they suggest that large-scale mining's dominance is largely pyrrhic and uncertain. Though the world continues to extract and consume its stock of ores and fossil fuels at furious rates, people and governments now regularly challenge these same forces, leaving prediction of extraction's future a fool's game.

These chapters highlight these issues via explanation and description of different structures, processes and contestations. Still, the chapters are nevertheless unified in what they imply about a future Earth in the throes of what Pijpers and Eriksen (2019: 9–10) term overheating. For mining, overheating consists of adoption of new and remarkably (over-) productive transformational technologies, and extraction's expansion out of the advanced economies and explosively into and within developing nations. Such expansion furthered an extraction dependency previously unknown in the political economy of developing regions, contributing to intensive political change and economic dislocation. Furthermore, the time dimensions of mining cycles, from planning to execution to consumption and often to termination, are vastly speeded up in the overheated world.

Overheating and related resource overuse and contestation are clear throughout the book. The most extreme case concerns the threats of state-enabled uranium mining to Navajo land and life and people's organizational and cultural resistance against uranium's disastrous toll (Mesbah). Similarly, the conference halls of Andalusia (Buu-Sao) are awash in contradiction as the well-connected offer paeans to sustainability to both preserve their capital and reputations, though questions remain as to their sincerity. Meanwhile, Kolwezi Congolese miners (Geenen) contradictorily expose their health and lives to greater risk by the illegal tunnels and earthworks they dig beneath their huts to access precious cobalt. The petrol smugglers of Bolivia (Ødegaard) also are caught in an extractive overheating-based shell game. They resist the state to preserve their illegal livelihood gained from an extractive resource that, of global necessity, must be disbanded to reduce overheating but in actuality continues to expand its range to disastrous climatological effects. Contrarily, the transformation of Svalbard from coal field to tourist destination (to local misfortune) enabled by the oil wealth of the Norwegian state offers an exception that proves the rule of threat and contestation.

Peoples' histories and identities too are sources of struggle in an overheated world of intensified extraction. Among Death Valley Timbisha Shoshone (White), the loss of ancestral lands to mining interests, supported by complicit governments, suggests the global spread of mining at the expense of the powerless, often the indigenous. But the contemporary moment has also spawned a range of popular attempts to gain greater control over the lives and identities mining has threatened. The Shoshone ultimately reclaimed their land much as the descendants of slaves in Ouro Preto, Brazil (De Souza Santos) reclaimed their community and their sense of pride in their struggle over the use of the mining-related built environment. Finally, the fraught identity of the Italian Piedmont miners (Zanini), culturally subordinated to the autochthonous valley population, speaks to the struggle over mining globally, as intensified extraction consumes the world and the world begins to fight back.

Along with questions of inequality and contestation, another level of significance within this book – occasionally glaring (as in discussions of Svalbard climate change) though usually implicit – addresses the most critical question confronting global mining for now and in the future. This, of course, is the challenge to mining of human-induced climate change and other disjunctures of the Anthropocene environment. Together these transformations demand we reconsider this volume in a more climatological light. In that sense, the significance and effects of mining are especially transformed, scaled, disbursed and intensified in a context of changing climate (Odell, Bebbington and Frey 2018). Biermann et al. (2016: 347) suggest this is reflected in 'interdependencies, inequalities and disparities … that have important consequences for the governance challenge of the Anthropocene'. It is these issues that this Afterword largely considers.

Issues of Anthropocenic climate change[1] scarcely appeared at the Cagliari event for which the contributions to this volume were originally written, and the authors only obliquely address the environmental stresses of mining to the planet and people. But any discussion of global mining in contemporary times must bring climate change front and centre. Consequently, below, I want to depart somewhat from strict consideration of this volume, to focus on the relation of Anthropocenic climate change and mining,[2] focusing first on how mining contributes to change in the physical planet and biological qualities of human communities. Subsequently, I consider how overheated mining disrupts a number of other forces contributing to planetary change. And finally, I highlight the importance of narrative and heritage in the mining sector, and how that articulates with climate change in the Anthropocene. By posing these questions and issues, I hope this Afterword will point out some new directions for the study of global mining, better integrating the changing Anthropocene climate circumstances while celebrating this volume for framing our understanding of the 'Global Life of Mines'.

Mining and Anthropocene Bodies

In the last few decades, before the global climate crisis became as recognized as it is and the pandemic upended the world as we know it, vast intensification has taken place in global mining. This exploding growth is thus both a critical condition and essential metaphor for the fevered, overheated state of the globe, manifested today in the ubiquity of climate crisis discourse and recognition of the role of climate change in the production of spin-offs like mass migration and pandemic diseases like Covid-19. Still, despite some adjustments here and there, and the infusion of sustainability rhetoric (Han Onn and Woodley 2014), neither impending climate catastrophe nor pandemic illness have much slowed the global overheating associated with mining, nor has scholarship yet effectively responded in terms of an expanded literature considering the relationship of mining and climate change (Odell, Bebbington and Frey 2018).

Evidence of mining is one key aspect documenting the realities, nature and conditions of the Anthropocene geological epoch. Dutch meteorologist Paul Crutzen (2002), along with biologist Eugene Stoermer, proclaimed the Anthropocene's emergence in the early twenty-first century. They saw the prevalence of carbon in recent geological strata as a chief diagnostic of this human-centred age. Carbon evidence clearly implies the importance of mining and combustion of extracted products as central to our troubling environmental reality. Mining's role in the production of Anthropocene conditions especially illustrates how humanity itself has 'become a geological agent' (Jonsson 2015: 55) and demands that modern climate science increasingly consider the role and meaning of extraction for the changing Earth.

Going forward, the implications of this are both glaring and troubling. In the first place, mining does curious things to the body of the Earth itself. Thus, recent images of diverse mining sites taken from space by remote sensing (Fisher and Hirschfeld 2020) document in graphic detail the effects of mining technologies, writ large, on the Earth's appearance. These images show the extent to which resource extraction changes the very nature of geological reality, and how such changes affect human actualities. These images are striking, both for their frightening beauty as well as for the destruction they document in particular micro-environments. In the Madre de Dios River basin of the Peruvian Amazon, deforestation caused by illegal gold mining is reminiscent of skin blotches and scales associated with dermatological disease. Similarly, the Bingham Canyon Mine in Utah seen from space is like the whorl of a giant fingerprint, metaphorically testifying to our imprint on our planetary home. Mining is also a chief cause of bioturbation, the forced rearrangement of soil morphology and sediments. Extraction is remarkably destructive of lithic structures and the chemistry of geological formations (Opperman 2018: 5) and even responsible for the creation of a number of new forms of minerals

(Grew 2017). Whether or not these minerals are useful or irrelevant for human use is beside the point. Their genesis clearly shows mining's role in disturbing and transforming the Earth, and thus creating challenges that must be dealt with directly by local communities.

More to the point, these geological images and effects affirm the corporal quality of extraction, such that effects on the body, both human and earthly, must remain a core focus of mining analysis. The physicality of mining, mine labour, mining's effects on bodies and the precarity of mining and its role in the life of communities has been occasionally celebrated, though more often reviled. George Orwell's (1937) descriptions of blackened coal miners of the English Midlands, June Nash's (1979) evocation of bent and broken bodies of Bolivian tin miners, or even my portrayal of mining's decline at the end of socialism and stresses on miner physiologies and psychologies (Kideckel 2008) attest to the importance of the corporal in the study of extraction.

Mesbah's chapter on uranium mining in New Mexico and White's on the removal of Native Americans in California for mining purposes also underscore mining's threat to human bodies. However, in all these instances, mining is only a cover for threats produced less by dust clouds and toxic metals and tailings than by systems of ownership and control. Extending the logic of these chapters, overheated extraction spawned by an interconnected and insatiable consumerist global culture produces scaled morbidities as problematic to collective human health and to our earthly home as any known. Perhaps this human predilection for transforming earthly geology and enabling disease vectors might encourage us to pair these chapters on 'The Global Life of Mines' with a companion set on 'The Mined Life of the Globe', elaborating the physical actualities facing the earth and its biological citizens that grow from overheated extraction.

Systems of Extraction and Sociopolitical Change

As the above suggests, overheated extraction is a vector for the climate and biological crises of the Anthropocene. The social and political conflicts and distortions these twin morbidities produce seem vast and enduring. Some of the issues of mining that exacerbate and/or reflect Anthropocenic climate change include expanded energy use and related growth of carbon emissions, deforestation, depletion and ruination of water resources, and as discussed above extensive land deformation. However, this is merely a list of the physical qualities of mining in relation to climate change. The challenge for us is to devise theoretical tools to explicate the processes by which climate and disease vectors produce these greater disruptions and to trace their effects on diverse sociopolitical conditions under which people live in the context of mining operations.

Previous attempts to trace these effects have been offered in past models of mining systems, such as stakeholder (Ballard and Banks 2003), assemblage (Roy Chowdhury and Lahiri-Dutt 2019) and minescape analysis (D'Angelo and Pijpers 2018). Stakeholder analysis examines the intersection of the diverse groups and individuals with interests in particular mining operations. Assemblage analysis defines any mining system as representing the intersection of a layer cake of different systems brought together by historical circumstances. Minescapes essentially combine the two, considering both the intersection of interest groups and the operation of historically different operating structures, legal regimes, technological processes and labour needs set within a changeful spatio-temporal context. But each of these view mining operations *sui generis* and pay little attention to exogenous factors (such as climate change and epidemic disease) that impinge on individual mining operations and the universe of mining as a whole.

Furthermore, mines are not whole systems in themselves. They are related and respond to a range of other conditions and pressures, and in so doing also influence other forces with which they interact. According to Ureta and Flores (2023) mines and the minerals they produce are located within a web of ecological relations such that to analyse mine governance or technologies or labour control or the particular uses of minerals in isolation from the entirety of their connections is misconstrued. This is especially the case in the conditions of the Anthropocene, where climatic influences appear large-scale and affect nearly every aspect of the ecological web in which particular mining operations, and extraction overall, are embedded. In responding to this changeful and challenged ecology, the human aspects of mining are upended and forced to adapt to mining's new conditions. This process is spearheaded by changed demands of operators, changed availability of mining occupations and changed environmental circumstances, not the least of which are climate threats and disease vectors.

Demographic variables are especially likely to be both affected and influential in the relation of mining to the crises of climate and biological integrity. Population change is especially notable in mining's feverish expansion and contraction of the last decades. As pointed out in the Cagliari discussions, and in some of this volume's contributions (e.g. De Souza Santos, White, Zanini, et al.), movement of people is a common and multidimensional feature of mining regions. On one hand, expansion of mining operations into diverse regions often results in large-scale dispossession and displacement of communities while also encouraging movement of new populations into regions to staff these enterprises. Climate change and epidemic disease scramble these processes. Echoing the expansion of European colonialism into indigenous territories, movement of mining into new zones not only brings in a new set of economic processes, new populations with which indigenous peoples must contend and new connections to diverse social and ecological processes, which, as such, also exposes populations to pernicious viruses and other disease vectors (Rodó et al. 2021: 576). This growth of mining then is a prime factor in creating the conditions for zoönotic 'spillover' (Quammen 2012).

The position of women is also likely to be affected by extraction-related climate and ecological change. The diverse circumstances of gender are not specifically addressed in this volume. However, women in mining communities live in more problematic conditions even before the pressures of climate change (Rolston 2014) and doubly so when considered in the context of the stresses and transformations related to climate change. First, climate change and related stresses can often break apart the complementarity of men's and women's roles in mining systems (Dublin 1998). Bakx and Gomez (2022) suggest how climate change produces increased competition for threatened resources, in which women are typically disadvantaged. Further, as mining interests shift production and location under pressure of change, those in control of these decisions will typically seek to deal with communities through their empowered groups and individuals (Le Meur and Banaré 2014), in which women are again disadvantaged. Finally, women's circumstances are also negatively affected by the reality that twenty-first-century mining is largely expanding into indigenous areas in the 'Global South' and thus within areas most intensely affected by climate change. All these factors make for a recipe of extreme stress on both women and climate, whose fortunes rise and fall with each other.

Ecological reasoning posits the expansion and contraction of populations and systems depending on competitive forces and transformations in resource availability. This essentially defines the circumstances of the world in conditions of climate change, translating into an instability of location, practice and governance. The ecology of mining systems in climate change and pandemic disease are especially evident in the expansion and contraction of mining. Some mining interests are hesitant to expand into previously inviting regions, while others' 'business as usual' approach ignores climate and disease threats and puts communities and workers at greater risk (Auciello 2020). Some closures are long in the works, like those in Andean Peru and Svalbard, Norway (Ødegaard, this volume). However, other mining operations only now have begun to respond to these twin forces, often in clumsy, contradictory ways. The EU's Integrated Strategy on Raw Materials cuts back extraction in the EU homeland but intensifies extraction in Latin America (Aguilar 2012: 4). Other localities hurried to close diverse mining complexes. South Africa shut down its entire mining industry in April 2019, furloughing 450,000 workers in a twenty-four-hour period. Mexico, meanwhile, shut down its mining operations, but then within a month, due to economic pressures, opened them again (Burmeister 2020). In the exception that proves the rule, the United States declared mining as 'Critical Manufacturing', thus preserving mining operations as part of an essential supply chain.

But whatever the particular response, the growth and decline of mining upends communities and regions, whether intended or not, thus leading to even greater struggles in and between communities, regions and nations over land use, production, organization of labour and the control of profit and prices. In response, some governments and mining companies use these crises to extend favourable mining concessions and repress community protests or to secure regulatory changes like reducing public oversight. Wildcatting and petty criminality expands among populations deprived of gainful mine employment or subjected to harsh and threatening labour. Among Congolese cobalt miners, the already rough, sometimes violent, competition to secure this valuable ore has taken on a new grim reality. Under the guise of pandemic social distancing, the Chinese companies that penetrated the industry restricted access to indigenous prospectors with regional government complicity. Locals were forced to engage in more problematic and dangerous behaviour (e.g. breaking and entering company mines, supporting child labour, digging deeper and unstable 'artisanal' mines), resulting in frequent deaths and injuries (Niarchos 2021). Similarly, the aggressive push of mining into previously protected parts of the Brazilian Amazon brought both extraction and pandemic to diverse indigenous populations (Watson 2020).

Growing from recognition of the climatological effects on and increasing costs of mining, and demands of governing agencies worldwide, the development of sustainable mining practices is also on the rise. Sustainability in extraction is a logical response to extreme climatological events, to changing availability of water resources and to effects on the land by unrestricted mining practices and the like, all of which ultimately affect the bottom line of mining operations. Despite the increased costs associated with sustainability practices, investors now encourage sustainability practices in a big way (Holger 2023). More to the point, sustainability in mining is also a response to the global outcry against human rights abuses in mining as well as to the degradation of the land coincidental to the use of extreme production methods, and the expansion of mining for rare earth metals, and related minerals like lithium, graphite and cobalt, to power our electronic devices (Early 2023).

At the same time, the pressures on developing countries to generate jobs and economic development tends to mean that sustainability practices are likely to receive short shrift while demands for social justice in these regions push mining companies to engage in intense 'greenwashing' campaigns to portray themselves as public benefactors. Thus, the spread of local empowerment in governance of mining regimes and the practice of sustainable mining remain uncertain and appear as much rhetorical as actual. Even the rhetoric of local control and sustainability is often directed to support factions or processes that further corporate dominance or greenwashing without actual change in practice (see Buu-Sao, this volume and Han Onn and Woodley 2014). Taken together then, these circumstances point to an uncertain future for sustainability in mining unless demands for these practices grow in response to international pressure and regulation amid the spread of climate crises.

Narrative and Heritage in Anthropocene Mining

The extent, uncertainty and threat of change in both climate and mining have prompted a heightened concern for narrative to help make sense of these troubling Anthropocenic conditions. David Wallace-Wells (2019) considers the developing climate crisis as the essential existential condition humanity must deal with for now and for the future. Our world is changing fast with uncertainties and destabilizing forces emerging under the impetus of climate change. Narrative is a way to make sense of this destabilizing world (Boccardi 2015) and to depict and process this transformation to possibly make better sense of the coming changes. As of yet, however, climate narratives are largely lacking (Moezzi, Janda and Rotmann 2017: 2) and particularly fail to address the so-called 'action gap' between policy narratives (what needs to be done) and implementation (how to do it) in climate action (Bushell et al. 2017). This is slowly beginning to change as advances in renewable energy and other climate initiatives shift the narrative from one of likely costs of the transition from a carbon-based economy to one where growth and riches follow successful climate remediation strategies (Dervis 2021).

Narratives of mining can make a contribution here. Given its time-depth, ubiquity, importance and dangers, mining has long been one of the major sources of narrative richness in labour and economic folklore. As Le Meur and Banaré suggest, 'The practice of mining is inseparable from its narrativization; of a non-mining past, a glorified or difficult present, a future characterized by extreme uncertainty at times and considerable certainty at others' (2014: 7). As such, the development of narrative approaches speaks of how mining was and how mining changes can set the pace for climate knowledge and action.

However, the contribution of mining to narrativize the climate crisis has not yet had an effect. In fact, mining narratives often encourage conflict and contradiction, thus avoiding grappling with the climate task. Mining narratives are often Manichaean, as mining conditions appear often equally so.[3] Thus, coal narratives pose how mining practices are world-destructive or, conversely, how coal combustion and so-called cheap heat and light enable satisfaction of human needs, thus restricting the will of polities to reform extraction governance and practice (McGowan 2020).[4] Such narratives often counterpose before and after circumstances, like an idealized or troubled non-mining past with a difficult or glorified present. The more tendentious the circumstances, the more aggressive master narratives about extraction practices, directions and significance.

Such narratives may emphasize how extraction serves development and modernization or despoliation and danger, as did Romanian socialist rhetorical claims of how 'coal made the nation' and 'kept it free and independent'. This is tantamount to fighting with words, like oppositional narratives that recently blossomed among contending sides struggling over coal's future in Appalachia. Executives regularly tout how coal provides low-cost, accessible and dependable energy to support living standards, especially for the world's poor. The phrase 'coal keeps the lights on' expresses this idea shorthand. But activists concentrate their rhetoric on a notion that 'coal is killing us'. By virtue of their extreme positioning, narratives breathe life into extraction mythologies. Those arguing over extraction's effects, whether positive or negative, often claim rampant extraction and post-extraction to press their case (see Sierra Club 2018). Expressions of anger about or support for extraction thus link narratives of past and present with doom or salvation, showing the ambiguity climate narratives face generally.

Narrative is thus still an imperfect vehicle to build support to address the growing climate emergency. Other 'soft' methods of explanation and instruction, such as heritage initiatives, are also not yet fully active in the climate challenge. But merely out of self-interest, mining heritage sites ought to begin to seriously address climate issues. It goes without saying that heritage institutions have been greatly disrupted, if not changed irrevocably by the pandemic and the growing climatological challenges of the Anthropocene. Weakened economies, persisting disease and climatological emergencies continually demand fiscal resources to support health and climate interventions, leaving fewer resources for heritage initiatives. Like so many other social and cultural desiderata, many heritage sites, practices and plans have been suspended and/or decimated by the fiscal and demographic challenges of coping with the pandemic and other climate-related emergencies.

Consequently, though challenged, heritage interpretation must begin to play an active role to explain the relationship of heritage resources to the circumstances of climate change. Right now, however, this is an imaginary project. Just as Michael Taussig (2004) mentally designed his idealized cocaine museum to manifest critical social, political and historical conditions that gave rise to the cocaine industry and trade in South America, so too must mining heritage institutions now and in the future develop an agenda that speaks to the significance of coal and other minerals and mining practices in producing the conditions behind the climate emergency. But they have yet to do so, as there is a disconnect between the macro-level – that is, global – provenance and significance of climate disturbance and the very localized histories that most mine heritage institutions consider. Furthermore, the coal/mining museum would seem ill-situated to expose the element that made their existence possible. Their missions tend to limit their activity. Finally, mining museums are often funded, in whole or in part, by mining interests, and are therefore highly unlikely antagonists to their benefactors. It is only where those financial interests are coal/mining-sceptical can climate change breakthrough as a topic of interest.

The Zeche Zollverein Coal Mine Industrial Complex in Essen, Germany, in the heart of the Ruhr former coal mining and industrial region, once the largest coal mine in the world, and a renowned UNESCO world heritage site, exemplifies a museum programme not beholden to coal interests. Zollverein, sponsored and funded by the North Rhein-Westphalia state Foundation, has responded to Germany's programme of *Energiewende* (the energy turnaround) by incorporating a detailed and sophisticated climate change programme around the idea of 'Bye-bye coal'.[5] Furthermore, throughout the pandemic, the site was host to outdoor concerts and festivals with mining and industrial themes to overcome the drag on culture that the pandemic represented, but its climate change mission is now front and centre.

Compared to Zollverein, the Phillips-Sprague model Mine and Museum in Beckley, West Virginia, restricts its exhibits and their interpretation to the history and political-economy of the local mining experience. Like Zollverein, the Beckley complex is publicly owned (by the city of Beckley), and its director is technically employed by the county Chamber of Commerce. Also, like Zollverein, the complex has mining exhibits – those focused on local history – and is engaged in a range of new outreach practices, exhibits and activities through the pandemic. The director in an interview recognized the climate emergency, how it affects the West Virginia environment and the role this plays in the pressures on coal, and the critical need for climate education. Unlike Zollverein, however, her governing authority depends on the coal economy and is supported in some of its efforts by the ladies auxiliary of the Friends of Coal organization. Though the director claims that the local coal interests have no say whatsoever over the museum's programme, she also says that people come to the museum to learn about coal mining, and the model coal train ride through the defunct Sprague mine is the highlight of people's visits, and she cannot imagine changing that.

A third example of the failure to include a climate perspective in coal mining heritage concerns the Petrila Mine in Romania's Jiu Valley, the most significant coal region in the country. Here a mine heritage complex opened in 2017, but these efforts have stalled and, in many ways, retrenched due to local governmental policies exacerbated by the pandemic. The mine itself was initially closed in 2015 and slated to be fully demolished. However, a number of local activists, joined by students and faculty from universities outside the region, engaged in a protracted battle with authorities, finally resulting in the preservation of a variety of buildings on the mine grounds that were turned into a museum complex. These included theatre space, a photo gallery, a museum devoted to mine rescue teams and a permanent art exhibition in what was once the miners' shower and dressing room. The refurbished complex had drawn great publicity, aided by the production of an avant-garde video, Planeta Petrila, that aired on HBO Europe, recounting the battle to refurbish the mine for heritage purposes. Given its critical origins and organization, a section was planned to include a significant exhibit on coal in the regional environment and in climate change but was forestalled due to the pandemic.

Much like the Beckley West Virginia Model Mine Museum, the Petrila complex was forced to close throughout most of the pandemic, and only since Summer 2021 has it begun to allow a restricted number of visitors. Nonetheless, despite the positive notoriety the refurbished mine space generated for the community, the local authorities, the formal owners of the land and buildings, sought to extract additional monies from the property and have blanched at extending the museum space. Unrelated to the pandemic, the Petrila Group has applied for recognition as a site on the European Route of Industrial Heritage (ERIH), a climate-sensitive project, but the ownership of the land and buildings by the local government has prevented that from being accepted. As one of the local Petrila activists said about that and about their plans for climate action, 'the worst pandemic Petrila suffers is the "pandemic of politics' ".

Thus, as the above suggests, the circumstances and transformations of the pandemic as well as the ongoing changes and challenges of the Anthropocene more generally demand we enlarge our understanding of heritage and extend the kinds of activities and processes that fall under the rubric of heritage development. Mining heritage must consider the diverse environments and geologic strata and flora and fauna that extraction disrupted. Following a path of mineral production, from extraction, through processing, to consumption, abandonment or recycling, might be a successful tool for any mine heritage institution. Similarly, examining the social relations and settlement structures of extractive communities and how they are transformed under climate change must be included in the mix.

Conclusions: Framing and Addressing the New 'Curse' of Extraction

Global mining is caught in the jaws of a dilemma. Growth in the sector, focused on accessing rare and precious metals and minerals to support the expanded use of electronic devices, vehicles and renewable energy systems, is unparalleled and seemingly without bottom. But expansion of mining intensifies all the issues and struggles defined in these chapters, and then some. Manipulation of land ownership; stolen resources and identities; exploitation and precarity of labour; expanded use of toxic substances; population transformations and discontinuities; and unfulfilled demands for genuine sustainability practices swirl about an industry recognized as one of the chief sources of climate change and planetary threat. To top it off, the changes in mining, including those inspired by the pandemic, have rearranged the web of extraction ecologies and interests to such an extent that the sector's growth is imperilled by its success while human populations and earthly circumstances are imperilled by its growth.

This set of contributions, ranging wide across the globe, point out this critical contradiction and imply the necessity of action to counter if not reverse these trends. In their introduction, Pusceddu and Zerilli note how this volume represents an early effort to examine the circumstances of the global mining sector in a systematic and interconnected way. This concern for the interconnections of mining across the globe – how growth in one area parallels decline in another; how calls for sustainability produce hidden practices of over-exploitation; how the search for minerals connected to climate remediation produces climate destructive practices elsewhere; how the celebration of heritage forecloses challenge to centuries of ruin – appears to be a new 'curse' of extraction updated for Anthropocene circumstances. Thus, this book offers a leading-edge discussion not only directed towards the mining sector in itself but quite overtly to address the larger implications of extraction for our threatened age. What does a call to action look like in the present global circumstances of 'accursed' mining? In fact, there is much that can be done in a diversity of ways, large and small, though whether or not these changes will be enabled is another question indeed. Perhaps the most significant of changes might be made in governance practices, to encourage a greater role for community voices at every level. This would imply having concern for the demographic instability introduced in mining regimes – for the physical needs of migrant and host populations alike.

New forms of governance would also require getting serious about inequality and sustainability and demand a meaningful partnership of capital, labour, nongovernmental groups and government at every level, to address the challenges of extraction in the context of climate change. Above all, to modify global mining demands recognition and making visible its role in the problem of changing climate and encouraging the systematic contribution of industry to the large-scale solutions demanded for these extraordinary times. This particularly implies a role for mining heritage institutions to develop a heightened concern for documenting the role of mining in the past production of climate-threatening practices and products.

In fact, much of this pushback against mining's overreaching (and overheated) policies is already happening as virtually all the contributions to this collection attest. The fight of Native American groups like Navajo and Shoshone has wrested their land back from Federal government licensees and private mining concessions. Congolese cobalt miners and Bolivian petrol smugglers resist through illegal labour. Brazilian slave descendants and Piedmontese miners resist through identity politics. And even powerful Norwegian tourism officials and Andalusian mine executives feel the pressures of challenges to mining and respond by offering concessions. These small incidents of pushback, when multiplied hundreds of times across the globe, provide some sense that all is not lost and that mining can ultimately be reined in for the good of the planet and its people.

But maybe all this is merely 'whistling past the graveyard' of our social, political and demographic realities in the face of impending climate catastrophe and epidemic disease. Nonetheless, even as we dawdle and temporize, the Anthropocene rumbles forward like some unstoppable glacier, rolling all up in its path. It behoves us to look with open eyes at the advancing massif, and at the extraction practices that fuel its growth, lest we too are ground to climate-induced rubble.

David A. Kideckel is professor emeritus of cultural anthropology at Central Connecticut State University, USA. He was founding chair of the Anthropology Department and originator/ occasional director of International and Area Studies. His last book, *Getting By in Post-Socialist Romania: Labor, the Body, and Working-Class Culture* (2008, Indiana U. Press), with a companion video, *Days of the Miners: Life and Death of a Working Class Culture,* concerned changing conditions and physical perceptions among workers after socialism.

Notes

1. The definition and reality of the Anthropocene are of some debate (Biermann et al. 2016; Skillington 2015). However, I find these discussions somewhat facile. The reality of climate change is upon us and likely to last and intensify from this day forward (Wallace-Wells 2019). How to name, date the origins of and define this epoch is really the least of our concerns. Instead, the operative questions are clearly 'what shall we do about this?' and for this Afterword, how does extraction influence the production of human-induced climate change and how might mining contribute to its amelioration?
2. In this Afterword, when I speak of mining I largely imply large-scale mining, mainly of corporate provenance.
3. The Manichean nature of mining was especially expressed by the idea of 'The Curse of Mining', where mining creates jobs and value but also dangerous conditions, and social malaise (Frankel 2010; Pijpers and Eriksen 2019).
4. Ironically, this idea was used by coal producers to justify their attempt to be excused from contributing to a fund to assist miners affected with Black Lung.
5. Energy turnaround notwithstanding, the pressures on the German economy of the Russian war against Ukraine and the rising cost of natural gas has again turned the German economy back towards coal. Over the last two years, Germany's use of coal for electricity generation has increased (Wall Street Journal 2023).

References

Aguilar, Carlos. 2012. *Transitions Toward Post-Extractive Societies in Latin America*. Dublin: Comhlámh.

Auciello, Benjamin H. 2020. 'Pandemic Profiteering: How Mining Companies Seek to Benefit from Covid-19'. *Earthworks Earthblog*. Retrieved 12 July 2023 from https://www.earthworks.org/blog/pandemic-profiteering-how-mining-companies-seek-to-benefit-from-Covid-19/.

Bakx, Jara, and Blanca Racionero Gomez. 2022. 'Gender Equality to Safeguard the Environment: How Women Can Improve the Environmental Performance of the Minerals Sector'. *Levin Sources*. Retrieved 20 September 2023 from https://www.levinsources.com/knowledge-centre/insights/gender-equality-safeguard-environment-mining.

Ballard, Chris, and Glenn Banks. 2003. 'Resource Wars: The Anthropology of Mining', *Annual Review of Anthropology* 32: 287–313.

Biermann, Frank et al. 2016. 'Down to Earth: Contextualising the Anthropocene', *Global Environmental Change* 39: 341–50.

Boccardi, Giovanni. 2015. 'From Mitigation to Adaptation: A New Heritage Paradigm for the Anthropocene', in Marie-Theres Albert (ed.), *Perceptions of Sustainability in Heritage Studies*. Berlin: Walter de Gruyter, pp. 87–98.

Burmeister, Tim. 2020. 'Mining during a Pandemic'. *Elko (Nevada) Daily Free Press*. Retrieved 12 July 2023 from https://elkodaily.com/mining/mining-during-a-pandemic/article_65a8da40-b45f-57b8-8600-9f44c93e6dc9.html.

Bushell, Simon, Géraldin Satre Buisson, Mark Workman and Thomas Colley. 2017. 'Strategic Narratives in Climate Change: Towards a Unifying Narrative to Address the Action Gap on Climate Change', *Energy Research & Social Science* 28: 39–49.

Crutzen, Paul J. 2002. 'Geology of Mankind', *Nature* 415: 23.

D'Angelo, Lorenzo, and Robert J. Pijpers. 2018. 'Mining Temporalities: An Overview', *The Extractive Industries and Society* 5(2): 215–22.

Dervis, Kemal. 2021. 'The New Climate Narrative'. *Brookings*. Retrieved 13 September 2023 from https://www.brookings.edu/articles/the-new-climate-narrative/.

Dublin, Thomas. 1998. *When the Mines Closed: Stories of Struggles in Hard Times*. New York: Cornell.

Early, Catherine. 2023. 'After Decades of Controversy, Can the Mining Industry Come Clean?' *Reuters Sustainable Switch*. Retrieved 25 September 2023 from https://www.reuters.com/sustainability/climate-energy/after-decades-controversy-can-mining-industry-come-clean-2023-05-09/.

Fisher, Richard, and Javier Hirschfeld. 2020. 'The Scarred Landscapes Created by Humanity's Material Thirst'. *BBC Future*. Retrieved 12 July 2023 from https://www.bbc.com/future/article/20201117-mining-and-anthropocene-landscapes.

Frankel, Jeffrey A. 2010. 'The Natural Resource Curse: A Survey'. Working paper 15836. Cambridge: *National Bureau of Economic Research*. Retrieved 1 November 2019 from http://www.nber.org/papers/w15836.

Grew, Edward. 2017. 'The Anthropocene Epoch: How Mining Spurs Rapid Mineral Diversification', *Mining Technology*, 12 June. Retrieved 12 July 2023 from https://www.mining-technology.com/features/featurethe-anthropocene-epoch-how-mining-spurs-rapid-mineral-diversification-5839093/.

Han Onn, Adam, and Alan Woodley. 2014. 'A Discourse Analysis on How the Sustainability Agenda is Defined Within the Mining Industry', *Journal of Cleaner Production* 84: 116–127. https://www.sciencedirect.com/science/article/pii/S0959652614003199.

Holger, Dieter. 2023. 'Investors Push Mining Companies to Adopt Better Sustainability Standards'. *Wall Street Journal Sustainable Business*. Retrieved 25 September 2023 from https://www.wsj.com/articles/investors-push-mining-companies-to-adopt-better-sustainability-standards-11674644471.

Jonsson, Fredrik Albritton. 2015. 'Anthropocene Blues: Abundance, Energy', *RCC Perspectives* 2: 55–64.

Kideckel, David A. 2008. *Getting By in Postsocialist Romania: Labor, the Body, and Working-Class Culture*. Bloomington, IN: Indiana University Press.

Le Meur, Pierre-Yves, and Eddy Banaré. 2014. 'History and Stories: The Politics and Poetics of Mining Narratives in the South Pacific', *Journal de la Société des Océanistes* 138–39: 5–22.

McGowan, Elizabeth. 2020. 'Coal's Collapse under COVID-19 Adds Urgency to Just Transition Movement'. *Energy News Network*. Retrieved 12 July 2023 from https://energynews.us/2020/06/08/coals-collapse-under-covid-19-adds-urgency-to-just-transition-movement/.

Moezzi, Mithra, Kathryn B. Janda and Sea Rotmann. 2017. 'Using Stories, Narratives, and Storytelling in Energy and Climate Change Research', *Energy Research & Social Science* 31: 1–10.

Nash, June. 1979. *We Eat the Mines and the Mines Eat Us: Dependency and Exploitation in Bolivian Tin Mines*. New York: Columbia University Press.

Niarchos, Nicolas. 2021. 'Buried Dreams: The Dark Side of Congo's Cobalt Rush', *The New Yorker*, 31 May 2021.

Odell, Scott D., Anthony Bebbington and Karen E. Frey. 2018. 'Mining and Climate Change: A Review and Framework for Analysis', *The Extractive Industries and Society* 5(1): 201–14.

Opperman, Serpil. 2018. 'The Scale of the Anthropocene: Material Ecocritical Reflections', *Mosaic: An Interdisciplinary Critical Journal* 51(3): 1–17.

Orwell, George. 1937. *The Road to Wigan Pier*. London: Victor Gollancz.

Pijpers, Robert J., and Thomas Hylland Eriksen (eds). 2019. *Mining Encounters: Extractive Industries in an Overheated World*. London: Pluto Press.
Quammen, David. 2012. *Spillover: Animal Infections and the Next Human Pandemic*. New York: Norton.
Rodó, Xavier, Adrià San-José, Karin Kirchgatter and Leonardo López. 2021. 'Changing Climate and the COVID-19 Pandemic: More Than Just Heads or Tails', *Nature Medicine* 27: 576–84. https://www.nature.com/articles/s41591-021-01303-y.pdf.
Rolston, Jessica S. 2014. *Mining Coal and Undermining Gender: Rhythms of Work and Family in the American West*. New Brunswick: Rutgers University Press.
Roy Chowdhury, Arnab, and Kuntala Lahiri-Dutt. 2019. 'Hirashasan: Governing Diamonds in Central India', *Geoforum* 105: 32–42.
Sierra Club. 2018. 'Beyond Coal'. Retrieved 25 October 2019 from https://content.sierraclub.org/coal/about-the-campaign.
Skillington, Tracey. 2015. 'Theorising the Anthropocene', *European Journal of Social Theory* 18(3): 229–35.
Taussig, Michael. 2004. *My Cocaine Museum*. Chicago: University of Chicago Press.
Ureta, Sebastián, and Patricio Flores. 2023. *Worlds of Gray and Green: Mineral Extraction as Ecological Practice*. Berkeley, CA: University of California Press.
Wall Street Journal. 2023. 'Coal Keeps Germany's Lights On'. Retrieved 15 September 2023 from https://www.wsj.com/articles/germany-coal-energy-electricity-renewables-4700b442.
Wallace-Wells, David. 2019. *The Uninhabitable Earth*. Penguin.
Watson, Katy. 2020. 'Coronavirus: Disease Meets Deforestation at Heart of Brazil's Amazon'. *BBC On-line*. Retrieved 12 July 2023 from https://www.bbc.com/news/world-latin-america-52502369.

Index

Printed in the USA
CPSIA information can be obtained
at www.ICGtesting.com
LVHW050724170724
785210LV00002B/180

9 781805 395911